"十三五"江苏省高等学校重点教材
编号：2020-2-224
国家"双高计划"建设院校
人工智能技术应用专业群课程改革系列教材

本书是"智能工厂"系列教材之一，是国家"双高计划"建设院校人工智能技术应用专业群课程改革成果。

MCU Practical Technology
单片机实用技术

主编　左亚旻　顾卫杰

北京理工大学出版社
BEIJING INSTITUTE OF TECHNOLOGY PRESS

内 容 简 介

本教材以 8051 为内核的单片机为例，介绍了 8051 系列单片机的基本原理，并使用 C 语言进行系统编程及项目应用。教材以项目化进行组织，以"智能家居控制系统"的设计为实践教学的总目标，共划分为 4 个项目，每个项目再细分为 2～3 个阶梯式任务。教材中融入了中国元素，介绍了中国芯片的发展史、国产芯片设计及生产的知名企业等内容，让留学生可以了解中国在全球半导体行业的地位。

本教材为立体化教材，除文字内容外，还配有电子教学资源。扫一扫书中的二维码即可查阅对应的视频、电子课件、习题答案等内容。

通过本教材的学习，学生能够掌握单片机的软硬件维护和开发应用，同时也能掌握电子信息和单片机的专业英语，成为国际化通用人才。

版权专有　侵权必究

图书在版编目（CIP）数据

单片机实用技术 = MCU Practical Technology：英文 / 左亚旻，顾卫杰主编. --北京：北京理工大学出版社，2021.9（2022.6 重印）
ISBN 978-7-5763-0441-1

Ⅰ. ①单… Ⅱ. ①左…②顾… Ⅲ. ①单片微型计算机–高等职业教育–教材–英文 Ⅳ. ①TP368.1

中国版本图书馆 CIP 数据核字（2021）第 200098 号

出版发行 /	北京理工大学出版社有限责任公司
社　　址 /	北京市海淀区中关村南大街 5 号
邮　　编 /	100081
电　　话 /	（010）68914775（总编室）
	（010）82562903（教材售后服务热线）
	（010）68944723（其他图书服务热线）
网　　址 /	http://www.bitpress.com.cn
经　　销 /	全国各地新华书店
印　　刷 /	三河市天利华印刷装订有限公司
开　　本 /	787 毫米×1092 毫米　1/16
印　　张 /	10.25
字　　数 /	213 千字
版　　次 /	2021 年 9 月第 1 版　2022 年 6 月第 2 次印刷
定　　价 /	36.00 元

责任编辑 / 王晓莉
文案编辑 / 辛丽莉
责任校对 / 周瑞红
责任印制 / 施胜娟

图书出现印装质量问题，请拨打售后服务热线，本社负责调换

Brief Introduction

This textbook takes 8051 MCU as the example, introduces the basic principles of 8051 series MCU, and uses C language for system programming and project application. The textbook is organized as a project, with the design of "smart home control system" as the general objective of practical teaching. It is divided into 4 projects, and each project is further subdivided into 2–3 step tasks. The textbook incorporates Chinese elements and introduces the history of China's chip industry, as well as the well-known domestic chip design and production companies, so that international students can learn about China's position in the global semiconductor industry.

This textbook has electronic teaching resources. Scan the QR codes in the textbook to check the corresponding teaching videos, lecture notes, quiz answers and other contents.

Through the study of this textbook, students can understand the MCU software and hardware maintenance and application development, at the same time can learn about the professional English of electronic information.

前言

　　常州机电职业技术学院单片机应用技术课程组的教师经过十余年的不断努力,将单片机课程逐步建设为"国家级精品课程""国家级精品资源共享课"以及"省级英文授课精品课程"。作为精品课程的建设成果之一,本教材使用英文编写,不断充实和完善项目课程的方案;不断优化和完善课程内容设置,在实践教学领域不断更新教学内容,增进校企合作"开放式"教学;不断更新和完善课程网站,丰富教学资源,是一本高质量的体现工学结合特色的英文教材。

　　通过本教材的学习,学生能够掌握单片机的软硬件维护和开发应用,同时也能掌握电子信息和单片机的专业英语,成为国际化通用人才。

本教材章节安排

　　本教材采用项目化的编写方式,以"智能家居控制系统"的设计为实践教学的目标,系统共分为 4 个项目,每个项目又分为 2~3 个阶梯式任务。课程的项目教学以单片机与接口技术在实际工程应用中的典型技术为背景,以单片机典型控制电路为载体,以 C 语言为软件设计程序,设置了 4 个项目共 10 个任务。

项目	任务
项目 1　家用灯光照明控制系统	任务 1　点亮 1 个 LED 发光二极管
	任务 2　8 个 LED 发光二极管的闪烁
	任务 3　按键控制 LED 发光二极管
项目 2　家用门禁报警控制系统设计	任务 1　独立按键门禁控制系统
	任务 2　矩阵键盘门禁控制系统设计
	任务 3　使用中断设计的独立按键门禁系统

续表

项目	任务
项目3　数字钟控制系统设计	任务1　简易秒表设计
	任务2　具有校时功能的24小时数字钟的设计
项目4　家用通信系统设计	任务1　通过RS232发送一条简单的信息
	任务2　通过RS232收发信息

本教材编写特点与创新

1. 本教材为一体化教材，除文字内容外，还配有电子教学资源。直接扫一扫书中的二维码即可查阅对应的一体化教学资源。

2. 本教材中融入了中国元素，介绍了中国芯片的发展史、国产芯片设计及生产的知名企业等内容，让留学生可以了解中国在全球半导体行业的地位。

3. 本教材的编程语言为C语言，采用"用到什么讲什么"的编写方式，将C语言的知识融入单片机技术中去讲解。

本教材的重点在于让学生学会使用单片机设计控制系统并能维护软硬件，而不是研究单片机的内部结构等理论知识。因此，在编写设计时，以项目化方式安排学习内容，以电子产品的开发步骤安排授课流程，课程偏重"实用"二字。

Preface

After more than ten years efforts, teachers of the MCU application technology curriculum group in Changzhou Vocational Institute of Mechatronic Technology have gradually developed the MCU course into "National Quality Course" "National Quality Resource Sharing Course" and "Provincial Quality Course Taught in English". As one of the achievements of the construction of high-quality courses, this textbook is written in English to constantly enrich and improve the program of the project curriculum. We constantly optimize and improve the course content setting, update the teaching contents in the field of practical teaching, and enhance the school-enterprise cooperation teaching. The course website is constantly updated and improved to enrich teaching resources. It is a high quality work-study combination textbook.

Through the study of this textbook, students will be able to master the software and hardware maintenance, development and application of MCU. At the same time, they can also master electronic information and MCU professional English.

Chapter arrangement

The textbook is written by projects, with the design of "intelligent home control system" as the goal of practical teaching. The system is divided into 4 projects, and each project is divided into 2~3 stepped tasks. The project teaching of the course takes the typical technology of MCU and interface technology in the practical engineering application as the background, takes the typical control circuit of MCU as the carrier, takes C language as the software design program, and sets up four projects with a total of 10 tasks.

Project	Task
Project 1　Household Lighting Control System Design	Task 1　One LED Basic Operation
	Task 2　Eight LEDs Basic Operation
	Task 3　Button Interface
Project 2　Household Access Control System Design	Task 1　Independent Button Access Control System Design
	Task 2　Matrix Keyboard Access Control System Design
	Task 3　Independent Button Access Control System Design with Interrupt
Project 3　Digital Clock Control System Design	Task 1　Simple Stopwatch Design
	Task 2　Design of 24 - hour Digital Clock with Timing Function
Project 4　Household Communication Control System Design	Task 1　Output a Simple Text Message from the RS232 Port
	Task 2　Input/Output Example Using the RS232 Port

Features and innovations

1. The textbook is equipped with electronic teaching resources in addition to text content. Scan the QR codes directly in the textbook, you can refer to the corresponding electronic teaching resources.

2. The textbook incorporates Chinese elements, introducing the history of China's chip development, domestic chip design and production of well-known enterprises, so that international students can understand China's position in the global semiconductor industry.

3. The programming part of this textbook uses C language, the writing method follows "teaching what you need", and the knowledge of C language is integrated into the MCU technology to explain.

The focus of this textbook is to let the students learn to use the MCU to design the control system and maintain the hardware and software, rather than study the internal structure of the MCU and other theoretical knowledge. Therefore, the textbook arranges the learning contents in a project-based way, the teaching process is arranged according to the development steps of electronic products, and the course focuses on "practical".

CONTENTS

Project 1　Household Lighting Control System Design ······················· 1

　Task 1　One LED Basic Operation ································· 1
　　Objectives ··· 1
　　Task Requirement ·· 1
　　1.1　MCU Application System ······································ 1
　　1.2　Create a New Project—the Use of Keil C51 Software ········ 4
　　Circuit Diagram ·· 15
　　Program Description ··· 17
　　Quiz ··· 17
　　Summary ··· 18
　Task 2　Eight LEDs Basic Operation ······························· 19
　　Objectives ··· 19
　　Task Requirement ·· 19
　　2.1　MCU Internal Resources ······································· 19
　　2.2　Minimum Microcontroller Configuration ····················· 22
　　2.3　MCU Memory Construction ···································· 25
　　2.4　Structure of a Microcontroller-based C Program ············ 31
　　2.5　The Basic of C Language: Data Types and Operators ······· 33
　　2.6　The Basic of C Language: Iteration Statements—for and while ······ 46
　　2.7　The Basic of C Language: Functions ·························· 50
　　Circuit Diagram ·· 53
　　Program Description ··· 53
　　Quiz ··· 55

Summary ·· 56
Task 3　Button Interface ··· 56
　　Objectives ·· 56
　　Task Requirement ··· 57
　　3.1　MCU Parallel I/O Port ·· 57
　　3.2　Use of Individual Buttons ·· 61
　　3.3　The Basic of C language: Selection Statement—if-else ·············· 62
　　Circuit Diagram ··· 65
　　Program Description ··· 66
　　Quiz ··· 67
　　Summary ··· 68
　　Extended Reading ··· 68

Project 2　Household Access Control System Design ··················· 70

Task 1　Independent Button Access Control System Design ················ 70
　　Objectives ·· 70
　　Task Requirement ··· 70
　　1.1　7-segment Display Driver ·· 71
　　1.2　Digital Tube Static Display ··· 75
　　1.3　The Basic of C Language: One Dimensional Array ·················· 75
　　1.4　The Basic of C Language: Selection Statement—Switch ··········· 79
　　Circuit Diagram ··· 83
　　Program Description ··· 84
　　Quiz ··· 85
　　Summary ··· 86
Task 2　Matrix Keyboard Access Control System Design ···················· 86
　　Objectives ·· 86
　　Task Requirement ··· 86
　　2.1　MCU and Matrix Keyboard Interface ··································· 86
　　Circuit Diagram ··· 89
　　Program Description ··· 89
　　Quiz ··· 92
　　Summary ··· 92
Task 3　Independent Button Access Control System Design with Interrupt ······ 93
　　Objectives ·· 93
　　Task Requirement ··· 93
　　3.1　MCU Interrupt System ·· 93

  3.2 MCU Interrupt Processing Function ········· 97
  Circuit Diagram ········· 98
  Program Description ········· 99
  Quiz ········· 99
  Summary ········· 100
  Extended Reading ········· 101

Project 3 Digital Clock Control System Design ········· 103

 Task 1 Simple Stopwatch Design ········· 103
  Objectives ········· 103
  Task Requirement ········· 103
  1.1 Timer/Counters' Structure and Working Mode ········· 103
  Circuit Diagram ········· 108
  Program Description ········· 109
  Quiz ········· 110
  Summary ········· 111
 Task 2 Design of 24 - hour Digital Clock with Timing Function ········· 111
  Objectives ········· 111
  Task Requirement ········· 111
  2.1 Timer/Counter Interrupts ········· 112
  2.2 Digital Tube Dynamic Display ········· 112
  Circuit Diagram ········· 113
  Program Description ········· 115
  Quiz ········· 118
  Summary ········· 119
  Extended Reading ········· 119

Project 4 Household Communication Control System Design ········· 122

 Task 1 Output a Simple Text Message from the RS232 Port ········· 122
  Objectives ········· 122
  Task Requirement ········· 122
  1.1 Serial Communication Basis and Serial Interface ········· 122
  1.2 RS232 Serial Communication ········· 129
  Circuit Diagram ········· 133
  Program Description ········· 134
  Quiz ········· 137
  Summary ········· 138

Task 2　Input/Output Example Using the RS232 Port ················ 138
　　Objectives ··· 138
　　Task Requirement ··· 138
　　Circuit Diagram ·· 139
　　Program Description ··· 140
　　Quiz ··· 143
　　Summary ·· 144
　　Extended Reading ··· 144

Glossary ·· 147

References ··· 152

Project 1

Household Lighting Control System Design

Task 1 One LED Basic Operation

Objectives

After completing this task, you should be able to
- understand what is MCU and be familiar with the use of Keil C51 software, through one LED basic operation.
- explain the structure of computer hardware.
- describe memory technologies.

Task Requirement

Use Protues and Keil C51 softwares to simulate the one LED basic operation. MCU controls LED's turning on and turning off.

1.1 MCU Application System

Scan the QR code to view the teaching note of MCU Introduction

Scan the QR code to watch the teaching video of MCU Introduction

Introduction

The term microcomputer is used to describe a system that includes a microprocessor, a program memory, a data memory, and input/output (I/O) ports. Some microcomputer systems include additional components such as timers, counters, analogue-to-digital converters and so on. Thus, a microcomputer system can be anything from a large computer system having hard disks, floppy disks and printers, to single chip computer systems.

In this textbook, we are going to consider only the type of microcomputers that consist of a single silicon chip. Such microcomputer systems are also called microcontrollers.

Microcontroller Evolution

First, microcontrollers were developed in the mid-1970s. These were basically calculator-based processors with small ROM program memories, very limited RAM data memories, and a handful of input/output ports. As silicon technology developed, more powerful, 8-bit microcontrollers were produced. In addition to their improved instruction sets, these microcontrollers included on-chip counters/timers, interrupt facilities, and improved I/O handling. On-chip memory capacity was still small and was not adequate for many applications. One of the most significant developments at this time was the availability of on-chip ultraviolet erasable EPROM memory. This simplified the product development time considerably and, for the first time, also allowed the use of microcontrollers in low-volume applications.

The 8051 family was introduced in the early 1980s by Intel. Since its introduction, the 8051 has been one of the most popular microcontrollers and has been second-sourced by many manufacturers. The 8051 currently has many different versions and some types including on-chip analogue-to-digital converters, a considerably large size of program and data memories, pulse-width modulation on outputs, and flash memories that can be erased and reprogrammed by electrical signals.

Microcontrollers have now moved into the 16-bit market. 16-bit microcontrollers are high-performance processors that find applications in real-time and compute intensive fields (e.g. in digital signal processing or real-time control). Some of the 16-bit microcontrollers include large amounts of program and data memories, multi-channel analogue-to-digital converters, a large number of I/O ports, several serial ports, high-speed arithmetic and logic operations, and a powerful instruction set with signal processing capabilities.

Microcontroller Architecture

The simplest microcontroller architecture consists of a microprocessor, a memory, and the input/output ports. The microprocessor consists of a central processing unit (CPU) and the control unit (CU).

CPU is the brain of a microprocessor and is where all of the arithmetic and logical operations are performed. The control unit controls the internal operations of the microprocessor and sends control signals to the other parts of the microprocessor to carry out the required instructions.

Memory is an important part of a microcomputer system. Depending upon the application, we can classify memories into two groups, program memory and data memory. Program memory stores all the program codes. This memory is usually a read-only memory (ROM). Other types of memories, e.g. EPROM and PEROM flash memories, are used for low-volume applications and during the period of the program development. Data memory is a read/write memory (RAM). In complex applications where there may be a need for large amounts of memories, it is possible to interface external memory chips to most microcontrollers.

Input/Output (I/O) ports allow external digital signals to be connected to the microcontroller. I/O ports are usually organized into groups of 8 bits and each group is given a name. For example, the 8051 microcontroller contains four 8-bit I/O ports named P0, P1, P2, and P3. On some microcontrollers, the direction of the I/O port lines are programmable so that different bits of a port can be programmed as inputs or outputs. Some microcontrollers (including the 8051 family) provide bi-directional I/O ports. Each I/O port line of such microcontrollers can be used as inputs and outputs. Some microcontrollers provide open-drain outputs where the output transistors are left floating (e.g. port P0 of the 8051 family). External pull-up resistors are normally used with such output port lines.

8051 Family

The 8051 family is a popular, industry standard 8-bit single chip microcomputer (microcontroller) family, manufactured by various companies with many different capabilities. The basic standard device which is the first member of the family, is the 8051, which is a 40-pin microcontroller. This basic device is now available in several configurations. The 80C51 is the low-power CMOS version of the family. The 8751 contains EPROM program memory, used mainly during the period of the development work. The 89C51 contains the flash programmable and erasable memory (PEROM) where the program memory can be reprogrammed without erasing the chip with ultraviolet light. The 8052 is an enhanced member of the family which contains more RAMs and also more timers/counters. There are many versions of the 40-pin family which contain one chip analogue-to-digital converters, pulse-width modulators, and so on. At the lower end of the 8051 family, we have the 20-pin microcontrollers which are code compatible with the 40-pin devices. The 20-pin devices have been manufactured for less complex applications where the I/O requirements are not very high and where less power is required (e.g. in portable applications). The AT89C1051 and AT89C2051 (manufactured by Atmel) are such microcontrollers, which are fully code compatible with the 8051 family and offer reduced power and less functionality. Table 1.1.1 gives a list of the characteristics of some popular members of the 8051 family.

Table 1.1.1 Some popular members of the 8051 family

Device	Program memory	Data memory	Timers/counters	I/O pins	Pin count
AT89C1051	1K flash	64 RAM	1	15	20
AT89C2051	2K flash	128 RAM	2	15	20
AT89C51	4K flash	128 RAM	2	32	40
AT89C52	8K flash	256 RAM	3	32	40
8051AH	4K ROM	128 RAM	2	32	40
87C51H	4K EPROM	128 RAM	2	32	40
8052AH	8K ROM	256 RAM	3	32	40
87C52	8K EPROM	256 RAM	3	32	40
87C54	16K EPROM	256 RAM	3	32	40
87C58	32K EPROM	256 RAM	3	32	40

In this textbook, all the projects are based upon the AT89C51 microcontroller. The reasons for choosing the AT89C51 are its low cost, low power consumption and powerful features.

1.2　Create a New Project—the Use of Keil C51 Software

Scan the QR code to view the teaching note of Creat a New Project

Scan the QR code to watch the teaching video of Creat a New Project

Keil C51 is the most popular software to develop 8051 microcontrollers. In this textbook, we describe specific features of the μVision user interfaces and how to interact with them. And we learn how to create projects, edit source files, compile, fix syntax errors, and generate executable codes. The Keil C51 is, for most developers, the easiest way to create embedded system programs.

To launch μVision, click the μVision icon on your desktop or select μVision from the

Start Menu.

Window Layout Concepts

You can set up your working environment in μVision at your discretion. Nevertheless, let us define three major screen areas. The definition will help you to understand future comments, illustrations, and instructions. (See Figure 1.1.1)

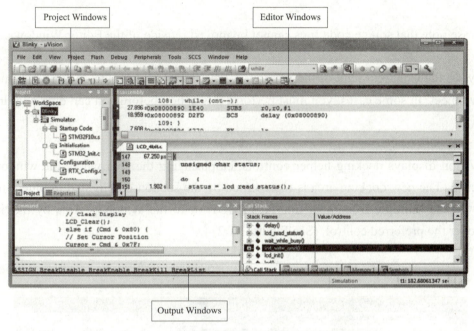

Figure 1.1.1   Window layout

The **Project Windows** area is the part of the screen in which, by default, the Project Window, Functions Window, Books Window, and Registers Window are displayed.

Within the **Editor Windows** area, you are able to change the source code, view performance and analysis information, and check the disassembly code.

The **Output Windows** area provides information related to debugging, memories, symbols, call stack, local variables, commands, browse information, and find in files results.

If, for any reason, you do not see a particular window and have tried displaying/hiding it several times, please invoke the default layout of μVision through the **Window—Reset Current Layout** Menu.

Positioning Windows

The µVision windows may be placed onto any area of the screen, even outside of the µVision frame, or to another physical screen. This only requires two steps.

- Click and hold the Title Bar of a window with the left mouse button.
- Drag the window to the preferred area, or onto the preferred control, and release the mouse button.

Please note that source code files cannot be moved outside of the Editor Windows.

Invoke the Context Menu of the window's Title Bar to change the docking attribute of a window object. In some cases, you must perform this action before you can drag and drop the window.

µVision displays docking helper controls, emphasizing the area where the window will be attached. The new docking area is represented by the section highlighted in blue. Snap the window to the Multiple Document Interface (MDI) or to the Windows Area by moving the mouse over the preferred control. (See Figure 1.1.2)

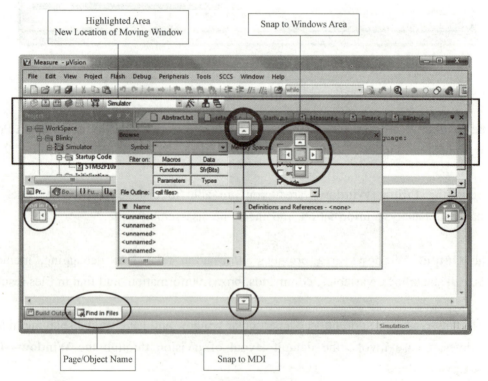

Figure 1.1.2   Positioning windows

Toolbars and Toolbar Icons

The μVision IDE incorporates several Toolbars with buttons for the most commonly used commands.

- The **File Toolbar** contains buttons for commands used to edit source files, to configure μVision, and to set the project specific options.
- The **Build Toolbar** contains buttons for commands used to build the project.
- The **Debug Toolbar** contains buttons for commands used in the debugger.

The File Toolbar is always available, while the Build Toolbar and Debug Toolbar will display in their contexts. In both Build Mode and Debug Mode, you have the options to display or hide the applicable Toolbars.

Creating Embedded Programs

μVision is a Windows application that encapsulates the Keil microcontroller development tools as well as several third-party utilities. μVision provides everything you need to start creating embedded programs quickly.

μVision includes an advanced editor, project manager, and make utility, which work together to ease your development efforts, decrease the learning curve, and help you to get started with creating embedded applications quickly.

There are several tasks involved in creating a new embedded project:

- Creating a Project File;
- Using the Project Windows;
- Creating Source Files;
- Adding Source Files to the Project;
- Using Targets, Groups, and Files;
- Setting Target Options, Groups Options, and File Options;
- Building the Project;
- Creating a HEX File.

The section provides a step-by-step tutorial that shows how to create an embedded project using the μVision IDE.

Creating a Project File

Creating a new μVision project requires just three steps:

1. Select the Project Folder and Project Filename;
2. Select the Target Microcontroller;
3. Copy the Startup Code to the Project Folder.

Selecting the Folder and Project Name

To create a new project file, select the **Project—New Project...** Menu. This opens a standard dialog that prompts you for the new project file name. It is a good practice to use a separate folder for each project. You may use the **Create New Folder** button in this dialog to create a new empty folder.

Select the preferred folder and enter the file name for the new project. μVision creates a new, empty project file with the specified name. The project contains a default target and a file group name, which you can view on the **Project Window**.

Selecting the Target Microcontroller

After you have selected the folder and decided upon a file name for the project, μVision asks you to choose a target microcontroller. This step is very important, since μVision customizes the tool settings, peripherals, and dialogs for that particular device.

The **Select Device** dialog box lists all the devices from the μVision **Device Database**.

You may invoke this screen through the **Project—Select Device for Target...** Menu in order to change the target later. (See Figure 1.1.3)

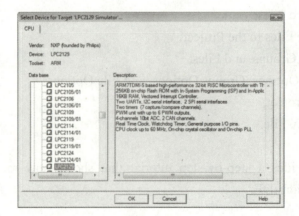

Figure 1.1.3   Select device for target

Copying the Startup Code

All embedded programs require some kind of microcontroller initialization or startup code that is dependent of the tool chain and hardware you will use. It is required to specify the starting configuration of your hardware.

All Keil tools include chip-specific startup code for most of the devices listed in the **Device Database**. Copy the startup code to your project folder and modify it there only. μVision automatically displays a dialog to copy the startup code into your project folder. Answer this question with **YES**. μVision will copy the startup code to your project folder and adds the startup file to the project. (See Figure 1.1.4)

The startup code files are delivered with embedded comments used by the configuration wizard to provide you with a GUI interface for startup configuration.

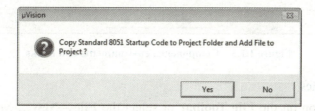

Figure 1.1.4　Copy standard 8051 startup code to project

Using the Project Windows

Once you have created a project file successfully, the **Project Window** shows the targets, groups, and files of your project. By default, the target name is set to **Target 1**, while the group's name is **Source Group 1**.

The file containing the startup code is added to the source group. Any file, the startup files included, may be moved to any other group you may define in future.

The **Books Window** which is also a part of the Project Windows, provides the Keil product manuals, data sheets, and programmer's guides for the selected microcontroller. Double-click a book to open it.

Right-click the **Books Window** to open its **Context Menu**. Choose **Manage Books...** to invoke the **Components, Environments and Books** dialog to modify the settings of the exiting manuals or add your own manuals to the list of books. (See Figure 1.1.5)

Later, while developing the program, you may use the **Functions Window** and

Templates Window as well.

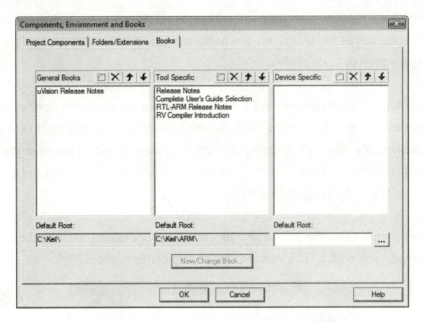

Figure 1.1.5   Components, environments and books

Creating Source Files

Use the button on the **File Toolbar** or select the **File—New...** Menu to create a new source file.

This action opens an empty **Editor Window** to enter your source code. μVision enables color syntax highlighting based on the file extension (after you have saved the file). To use this feature immediately, save the empty file with the desired extension prior to starting coding.

Save the new source file using the button on the **File Toolbar** or use the **File—Save** Menu.

Adding Source Files to the Project

After you have created and saved your source file, add it to the project. Files existing in the project folder, but not included in the current project structure, will not be compiled.

Right-click a file group in the **Project Window** and select **Add Files to Group** from the **Context Menu**. Then, select the source file or source files to be added. (See Figure 1.1.6)

A self-explanatory window will guide you through the steps of adding a file.

Project 1　Household Lighting Control System Design

Figure 1.1.6　Add files to group

Using Targets, Groups, and Files

The μVision's very flexible project management structure allows you to create more than one **Targets** for the same project.

A **Target** is a defined set of build options that assemble, compile, and link the included files in a specific way for a specific platform.

Multiple file groups may be added to a target and multiple files may be attached to the same file group.

You can define **multiple targets** for the same project as well.

You should customize the name of targets and groups to match your application structure and internal naming conventions. It is a good practice to create a separate file group for microcontroller configuration files.

🔧 Use the **Components, Environment, and Books…** dialog to manage your Targets, Groups, and Files configuration.

To change the name of a Target, Group, or File you may either double-click the desired item, or highlight the item and press F2. Change the name and click the **OK** button. Changes will be visible in the other windows as soon as this dialog has been closed. (See Figure 1.1.7)

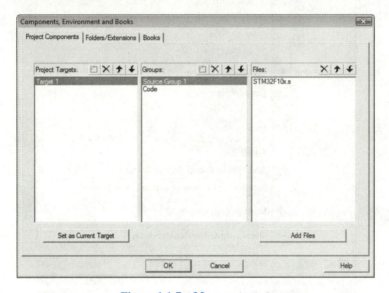

Figure 1.1.7　Manage target

Setting Target Options

🔖Open the Options for Target dialog from the Build Toolbar or from the Project Menu. (See Figure 1.1.8)

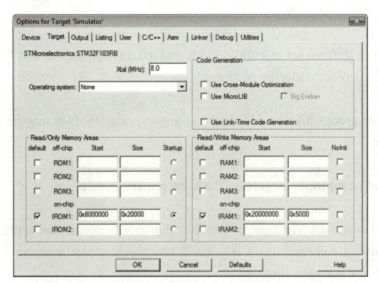

Figure 1.1.8　Options for target

Through this dialog, you can
- change the target device;

- set target options;
- and configure the development tools and utilities.

Normally, you do not have to make changes to the default settings in the **Target** and **Output** dialog.

The options available in the **Options for Target** dialogs depend on the microcontroller device selected. Of course, the available tabs and pages will change in accordance with the device selected and the target.

When switching between devices, the menu options are available as soon as the **OK** button in the **Device Selection** dialog has been clicked.

Building the Project

Several commands are available from the **Build Toolbar** or **Project** Menu to assemble, compile, and link the files of your project. Before any of these actions are executed, files are saved.

Translate File—Compiles or assembles the currently active source file.

Build Target—Compiles and assembles those files that have changed, then links the project.

Rebuild—Compiles and assembles all files, regardless whether they have changed or not, then links the project.

While assembling, compiling, and linking, μVision displays errors and warnings in the **Build Output Window**.

Highlight an error or warning and press **F1** to get help regarding that particular message. Double-click the message to jump to the source line that caused the error or warning. (See Figure 1.1.9)

μVision displays the message **0 Error(s), 0 Warning(s)** on successful completion of the build process. Though existing warnings do not prevent the program from running correctly, you should consider solving them to eliminate unwanted effects, such as time consumption, undesirable side effects, or any other actions not necessary for your program. (See Figure 1.1.10)

```
Build Output                                                                    ×
Build target 'Simulator'
assembling STM32F10x.s...
compiling Retarget.c...
compiling LCD_4bit.c...
compiling Serial.c...
compiling STM32_Init.c...
compiling Measure.c...
Measure.c(221): warning: #223-D: function "lfsfscd_clear" declared implicitly
compiling Getline.c...
compiling Mcommand.c...
linking...
.\Obj\Measure.axf: Error: L6218E: Undefined symbol lfsfscd_clear (referred from measure.o).
Target not created
```

Figure 1.1.9   Warning and error

```
Build Output                                                                    ×
Build target 'Simulator'
assembling STM32F10x.s...
compiling Retarget.c...
compiling LCD_4bit.c...
compiling Serial.c...
compiling STM32_Init.c...
compiling Measure.c...
compiling Getline.c...
compiling Mcommand.c...
linking...
Program Size: Code=8960 RO-data=1320 RW-data=52 ZI-data=1364
".\Obj\Measure.axf" - 0 Error(s), 0 Warning(s).
```

Figure 1.1.10   Build output

Creating a HEX File

Check the **Create HEX File** box under **Options for Target — Output**, and μVision will automatically create a HEX file during the build process.

Select the desired HEX format through the drop-down control to generate formatted HEX files, which are required on some Flash programming utilities. (See Figure 1.1.11)

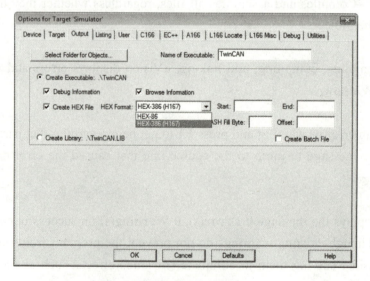

Figure 1.1.11   Options for target

Executing Code

μVision provides several ways to run your programs. You can

- instruct the program to run directly to the main C function. Set this option in the **Debug** tab of the **Options for Target** dialog;
  - select debugger commands from the **Debug** Menu or the **Debug Toolbar**;
  - enter debugger commands in the **Command Window**;
  - execute debugger commands from an initialization file.

Starting the Program

Select the **Run** command from the **Debug Toolbar** or **Debug** Menu or type **GO** in the **Command Window** to run the program.

Stopping the Program

Select **Stop** from the **Debug Toolbar** or from the **Debug** Menu or press the **Esc** key while in the **Command Window**.

Resetting the CPU

Select **Reset** from the **Debug Toolbar** or from the **Debug – Reset CPU** Menu or type **RESET** in the **Command Window** to reset the simulated CPU.

Single-Stepping

To step through the program and into function calls use the **Step** command from the **Debug Toolbar** or **Debug** Menu. Alternatively, you enter **TSTEP** in the **Command Window**, or press **F11**.

To step through the program and over function calls use the **Step Over** command from the **Debug Toolbar** or **Debug** Menu. Enter **PSTEP** in the **Command Window**, or press **F10**.

To step out of the current function, use the **Step Out** command from the **Debug Toolbar** or **Debug** Menu. Enter **OSTEP** in the **Command Window**, or press **Ctrl+F11**.

Circuit Diagram

Scan the QR code to view the teaching note of Build a Circuit and Download Program

Scan the QR code to watch the teaching video of Build a Circuit and Download Program

As is shown in Fig. 1.1.12, the circuit is extremely simple, consisting of the basic 89C51 based microcontroller and one LED connected to P1.0 of the microcontroller. Each microcontroller output pin can sink a maximum of 80µA and source up to 20mA. The manufacturers specify that the total source current of a port should not exceed 80mA. There are many different types of LED lights on the market, emitting red, green, amber, white, or yellow colours. Standard red LEDs require about 5 to 10mA to emit visible bright light. There are also low-current small LEDs, operating from as low as 1mA.

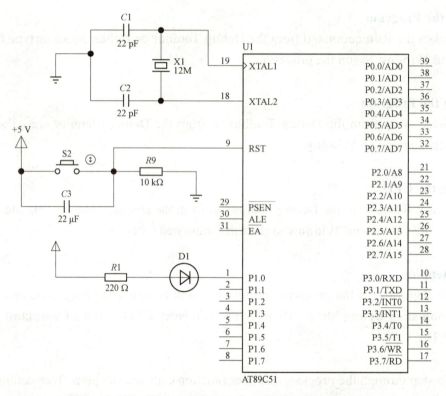

Figure 1.1.12 Circuit diagram of one LED basic operation

In Figure 1.1.12, the microcontroller outputs operate in current source mode where an LED is turned on if the corresponding output is at logic LOW level. The required value of the current limiting resistors can be calculated as follows:

$$R = \frac{V_s - V_f}{I_f}$$

where V_s is the supply voltage (+5V); V_f is the LED forward voltage drop (about 1.7V); and I_f is the LED forward current (3 to 20mA depending on the type of LED used). The required resistors will be:

$$R_{max} = \frac{(5-1.7) \text{ V}}{3 \text{ mA}} \cong 1.1 \text{ k}\Omega$$

$$R_{min} = \frac{(5-1.7) \text{ V}}{20 \text{ mA}} = 165 \text{ }\Omega$$

In this book, we often use the resistor with 220 Ω.

Program Description

```
/***********************************************************************
PROJECT: One LED Basic Operation
FILE: PROJ1_1.C
PROCESSOR: AT89C51
This project controls one LED turn on and turn off.
***********************************************************************/
#include <reg51.h>
sbit P1_0 = P1^0;              //variable P1_0 is assigned to bit 0 of Port 1

/* start of main program */
void main()
{
    unsigned int a;

    /*start of endless loop*/
    while(1)
    {
        for(a=0; a<50000; a++);     //delay
        P1_0 = 0;                   //turn off the LED
        for(a=0; a<50000; a++);     //delay
        P1_0 = 1;                   //turn on the LED
    }
}
```

Quiz

1. What is the MCU control core which complete operations and control functions? (   )
    A) CPU.
    B) RAM.
    C) ROM.

D) ALU.

2. MCU is essentially a (   ).
   A) Chip.
   B) Circuit board.
   C) Program.
   D) C51 software.

3. Which are the two parts of the MCU application system? (   )
   A) Arithmetic unit and the controller.
   B) Memory and register.
   C) Hardware system and software system.
   D) Input and output.

4. What are the two parts of a CPU? (   )
   A) Arithmetic unit and controller.
   B) Adders and registers.
   C) Arithmetic unit and adder.
   D) Arithmetic unit and decode.

5. Which of the following devices are external input devices? (   )
   A) LED.
   B) Switch.
   C) Keyboard.
   D) Mouse.

Scan the QR code to view the answer of quiz

Summary

Through the one LED basic operation, understand what is the MCU and be familiar with the use of Keil C51 software.

Project 1　Household Lighting Control System Design

Task 2　Eight LEDs Basic Operation

Objectives

After completing this task, you should be able to
- understand what is the MCU minimum microcontroller, and how to build a microcontroller-based C Program through the eight LEDs basic operation;
- explain the overall structure of a C language program;
- understand the basic data types and expressions of C language;
- perform basic I/O operations in the C language.

Task Requirement

8 LEDs are connected to Port 2. They are displayed by the following steps:

- All the LEDs are turned on for a few seconds;
- All the LEDs are turned off for a few seconds;
- The left 4 LEDs are turned on, while the right 4 LEDs are turned off for a few seconds;
- The right 4 LEDs are turned on, while the left 4 LEDs are turned off for a few seconds;
- Turn on one by one from left to right.

2.1　MCU Internal Resources

Scan the QR code to view the teaching note of MCU Internal Resources

Scan the QR code to watch the teaching video of MCU Internal Resources

Architecture of the 8051 Family

The 8051 is an 8-bit, low-power, high-performance microcontroller. There are a large number of devices in the 8051 family with similar architectures and each member of the family is downward compatible with each other. The basic 8051 microcontroller has the following features:

- 4 KB of program memory;
- 256 × 8 RAM data memory;

- 32 programmable I/O lines;
- Two 16-bit timer/counters;
- 6-source / 5-vector interrupt structure with priority levels;
- Programmable serial UART port;
- External memory interface;
- Standard 40-pin package.

Figure 1.2.1 shows 8051 family internal architecture.

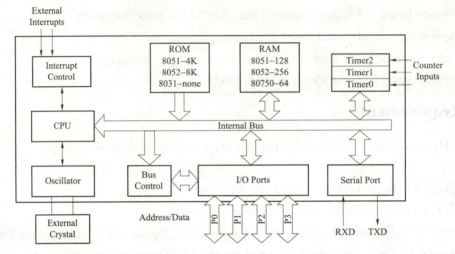

Figure 1.2.1   8051 Family internal architecture

Pin Configuration

The pin configuration of the standard 8051 microcontroller is shown in Figure1.2.2. Descriptions of the various pins are given below.

PIN 9 (RST): PIN 9 is the reset input. This input should normally be at logic 0. A reset is accomplished by holding the RST pin high for at least two machine cycles. Power-on reset is normally performed by connecting an external capacitor and a resistor to this pin.

PIN 18 (XTAL2) and PIN 19 (XTAL1): These pins are where an external crystal should be connected for the operation of the internal oscillator. Normally, two 33pF capacitors are connected with the crystal. A machine cycle is obtained by dividing the crystal frequency by 12. Thus, with a 12 MHz crystal, the machine cycle is 1 ms. Most machine instructions execute in one machine cycle.

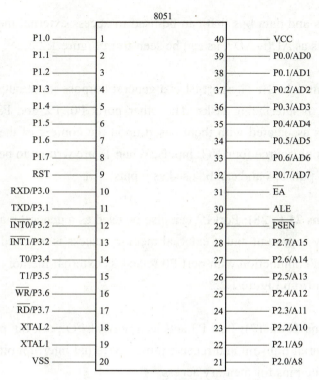

Figure 1.2.2 Pin diagram of the 8051 DIP

PIN 40 (VCC) and PIN 20 (VSS): Pins 40 and 20 are VCC and ground respectively. The 8051 chip needs +5V 500mA to function properly, although there are lower powered versions like the Atmel 2051 which is a scaled down version of the 8051 which runs on +3V.

PIN 29 ($\overline{PSEN}$): This is the program store enable pin on the 8051 devices. This pin is activated when the device is executing code from external memory.

PIN 30 (ALE): ALE means address latch enable, which is used when multiple memory chips are connected to the controller and only one of them needs to be selected.

PIN 31 ($\overline{EA}$): This is the external access enable pin on the standard 8051. $\overline{EA}$ should be connected to VCC for internal program executions. This pin also receives the programming voltage during programming.

There are four 8-bit ports: P0, P1, P2 and P3.

PORT P0 (Pins 32 to 39): Port P0 can be used as a general purpose 8-bit port when no external memory is present, but if external memory access is required then port P0 acts as a multiplexed address and data bus that can be used to access external memory in conjunction with port P2. P0 acts as AD0–AD7, as can be seen from Figure 1.2.2.

PORT P1 (Pins 1 to 8): The port P1 is a general purpose input/output port which can be used for a variety of interfacing tasks. The other ports P0, P2 and P3 have dual roles or additional functions associated with them based upon the context of their usage.The port P1 output buffers can sink/source four TTL inputs. When 1s are written to port P1, pins are pulled high by the internal pull-ups and can be used as inputs.

PORT P2 (Pins 21 to 28): Port P2 can also be used as a general purpose 8-bit port when no external memory is present, but if external memory access is required then port P2 will act as an address bus in conjunction with port P0 to access external memory. Port P2 acts as A8–A15, as can be seen from Figure 1.2.2.

PORT P3 (Pins 10 to 17): Port P3 acts as a normal I/O port, but port P3 has additional functions such as serial transmit and receive pins, 2 external interrupt pins, 2 external counter inputs, read and write pins for memory access.

2.2　Minimum Microcontroller Configuration

Scan the QR code to view the teaching note of Minimum Microcontroller Configuration

Scan the QR code to watch the teaching video of Minimum Microcontroller Configuration

The minimum microcontroller configurations of the 8051 based microcontroller systems are shown in Figure 1.2.3, Figure 1.2.5, Figure 1.2.6 and Figure 1.2.7.

Oscillator Circuit

The 8051 requires an external oscillator circuit. The oscillator circuit usually runs around 12 MHz, although the 8051 (depending on which specific model) is capable of running at a maximum of 40 MHz. Each machine cycle in the 8051 is 12 clock cycles, giving an effective cycle rate at 1 MHz (for a 12 MHz clock) to 3.33 MHz (for the maximum 40 MHz clock). The oscillator circuit generates the clock pulses so that all internal operations are synchronized.

The oscillator formed by the crystal, capacitors, and an on-chip inverter generates a pulse train at the frequency of the crystal, as is shown in Figure 1.2.4.

Figure 1.2.3 Oscillator circuit

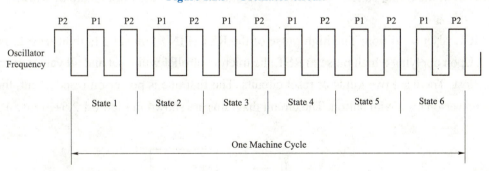

Figure 1.2.4 8051 timing

The clock frequency establishes the smallest interval of time within the microcontroller, called the pulse time. The smallest interval of time to accomplish a simple instruction, or part of a complex instruction, however, is the machine cycle. The machine cycle itself is made up of six states. A state is the basic time interval for discrete operations of the microcontroller such as fetching an opcode byte, decoding an opcode, executing an opcode, or writing a data byte. Two oscillator pulses define each state.

Program instructions may require one, two, or four machine cycles to be executed, depending on the type of instruction. Instructions are fetched and executed by the microcontroller automatically, beginning with the instruction located at ROM memory address 0000h at the time the microcontroller is first reset.

To calculate the time any particular instruction will take to be executed find the number of cycles. The time to execute that instruction is then found by multiplying C by 12 and dividing the product by the crystal frequency (f):

$$T=(C*12)/f$$

In conclusion, one machine cycle has 6 states. One state is 2 T-states. Therefore, one machine cycle is 12 T-states. Time to execute an instruction is found by multiplying C by 12 and dividing product by crystal frequency. For example, with a 12 MHz crystal, the machine cycle is 1 ms. An 11.059,2 MHz crystal, while seemingly an odd value, yields a cycle frequency of 921.6 KHz, which can be divided evenly by the standard communication baud rates of 19,200, 9,600, 4,800, 2,400, 1,200, and 300 Hz.

Reset Circuit

A reset circuit is a device used to restore a circuit to its original state. The 8051 microcontroller needs to reset when starting, in order to make the CPU and various parts of the system in the initial state of determination, and start from the initial state of work.

MCU reset condition: The high pulse of the RST (PIN 9) must be high at least 2 machine cycles. Upon applying a high pulse to RST, the microcontroller will reset and all values in registers will be lost. There are two kinds of reset circuits. The first one is power-on reset circuit, the other one is power-on reset with button. The circuit diagrams are shown in Figure 1.2.5–Figure 1.2.7.

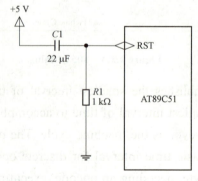

Figure 1.2.5   Power-on reset circuit

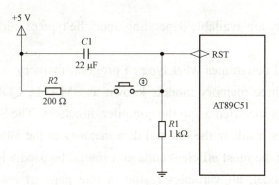

Figure 1.2.6　Electrical Level Reset Circuit with Button

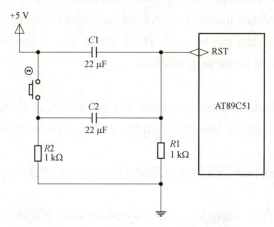

Figure 1.2.7　Pulse reset circuit with button

2.3　MCU Memory Construction

Scan the QR code to view the teaching note of MCU Memory Construction

Scan the QR code to watch the teaching video of MCU Memory Construction

8051 architecture supports both program (or code) and data memory areas. Program memory is read-only and it cannot be written into. Depending upon the type of processor used, different amounts of internal program memory are available. For example, 8051 provides 4 kbytes of internal program memory. The program memory can be increased by connecting additional external memory to the basic microcontroller. There may be up to 64 kbytes of program memory.

Data memory resides within the 8051 CPU and can be read from and written into. Up to

256 bytes of data memory are available depending upon the type of microcontroller used.

The memory model determines what type of program memory is to be used for a given application. There are three memory models known as SMALL, COMPACT, and LARGE, and the required model is specified using the compiler directives. The SMALL memory model is used if all the variables reside in the internal data memory of the 8051. This memory model generates the fastest and the most efficient code and should be used whenever possible. In the COMPACT memory model, all variables reside in one page of external data memory. A maximum of 256 bytes of variables can be used. This memory model is not as efficient as the SMALL model. In the LARGE memory model, all variables reside in external data memory. A maximum of 64 Kbytes of data can be used. The LARGE model generates more code than the other two models and thus it is not very efficient.

Compiling in the SMALL memory model always generates the fastest and the smallest code possible since accessing the internal memory is always faster than accessing any external memory.

The 8051 architecture provides four different physical memory regions.

DATA/IDATA memory includes a 256 bytes on-chip RAM with register banks and bit-addressable space that is used for fast variable accessing. Some devices provide an extended data (EDATA) space with up to 64 kbyte.

CODE memory consists of 64 kbyte ROM space used for program code and constants. The Keil linker supports code banking that allows you to expand the physical memory space. In extended variants, up to 16 mbyte ROM space is available.

XDATA memory has a 64 kbyte RAM space for off-chip peripheral and memory addressing. Today, most devices provide some on-chip RAM that is mapped into XDATA.

SFR and **IDATA** memory are located in the same address space but are accessed through different assembler instructions.

For extended devices, the memory layout provides a universal memory map that includes all 8051-memory types in a single 16 mbyte address region.

Program Counter and Data Pointer

The 8051 contains two 16-bit registers, the program counter (PC)and the data pointer

(DPTR). Each is used to hold the address of a byte in memory.

Program instruction bytes are fetched from locations in the memory that are addressed by the PC. Program ROM may be on the chip at addresses 0000h to 0FFFh, external to the chip for addresses that exceed 0FFFh, or totally external for all addresses from 0000h to FFFFh. The PC is automatically incremented after every instruction byte is fetched and may also be altered by certain instructions. The PC is the only register that does not have an internal address.

The DPTR register is made up of two 8-bit registers named DPH and DPL that are used to furnish memory addresses for internal and external code access and external data access. The DPTR is under the control of program instructions and can be specified by its 16-bit name, DPTR, or by each individual byte name, DPH and DPL. DPTR does not have a single internal address. DPH and DPL are each assigned an address.

A and B CPU Registers

The 8051 contains 34 general-purpose, or working registers. Two of these, registers A and B, comprise the mathematical core of the 8051 central processing unit(CPU). The other 32 are arranged as part of internal RAM in four banks, B0–B3, of eight registers each named $R0$ to $R7$.

The A (accumulator) register is the most versatile of the two CPU registers and is used for many operations, including addition, subtraction, integer multiplication and division, and Boolean bit manipulations. The A register is also used for all data transfers between the 8051 and any external memory. The B register is used with the A register for multiplication and division operations and has no other functions other than as a location where data may be stored.

Flags and the Program Status Word (PSW)

Flags are 1-bit registers provided to store the results of certain program instructions. Other instructions can test the condition of the flags and make decisions based upon the flag states. In order that the flags may be conveniently addressed, they are grouped inside the program status word (PSW) and the power control (PCON) registers.

The 8051 has four math flags that respond automatically to the outcomes of math operations and three general-purpose user fags that can be set to 1 or cleared to 0 by the programmer as desired. The math flags include carry (C), auxiliary carry(AC), overflow (OV), and parity (P). User flags are named F0, GF0, and GF1. They are general-purpose flags that may be used by the programmer to record some event in the program. Note that all of the flags can be set and cleared by the programmer at will. The math flags, however are also affected by math operations.

The program status word is shown in Table 1.2.1. The PSW contains the math flags user program flag F0, and the register select bits that identify which of the four general purpose register banks is currently in use by the program. The remaining two user flags GF0 and GF1, are stored in PCON.

Table 1.2.1 PSW program status word

Bit Address	0xD7	0xD6	0xD5	0xD4	0xD3	0xD2	0xD1	0xD0
Symbol	CY	AC	F0	RS1	RS0	OV	—	P

- CY (PSW.7): Carry flag.
- AC (PSW.6): Auxilliary carry flag; used for BCD arithmetic.
- F0 (PSW.5): User flag 0.
- RS1 (PSW.4): Register bank select bit 1.
- RS0 (PSW.3): Register bank select bit 0.
- OV (PSW.2): Overflow flag used in arithmetic instructions.
- — (PSW.1): Reserved for future use.
- P (PSW.0): Parity flag, shows parity of register A: 1=Odd Parity.

Internal Memory

A functioning computer must have memory for program code bytes, commonly in ROM and RAM memories for variable data that can be altered as the program runs. The 8051 has internal RAM and ROM memories for these functions. Additional memories can be added externally using suitable circuits.

Unlike microcontrollers with Von Neumann architectures, which can use a single memory address for either program code or data, but not for both. The 8051 has a Harvard architecture which uses the same address in different memories, for code and data. Internal circuitry accesses the correct memory based upon the nature of the operation in progress.

Internal RAM

The 128-byte internal RAM is organized into three distinct areas (See table 1.2.2):

- 32 bytes from address 00h to 1Fh that make up 32 working registers organized as four banks of eight registers each. The four register banks are numbered 0 to 3 and are made up of eight registers named R0 to R7. Each register can be addressed by name(when its bank is selected) or by its RAM address. Thus R0 of bank 3 is R0 (if bank 3 is currently selected) or address 18h (whether bank 3 is selected or not). Bits RS0 and RS1 in the PSW determine which bank of registers is currently in use at any time when the program is running. Register banks which are not selected can be used as general-purpose RAM. Bank 0 is selected upon reset.

- A bit-addressable area of 16 bytes occupies RAM byte addresses 20h to 2Fh, forming a total of 128 addressable bits. An addressable bit may be specified its bit address of 00h to 7Fh, or 8 bits may form any byte address from 20h to 2Fh. Thus, for example, bit address 4Fh is also bit 7 of byte address 29h. Addressable bits are useful when the program need only remember a binary event (switch on, light off, etc.). Internal RAM is in short supply as it is, so why use a byte when a bit will do.

- A general-purpose RAM area above the bit area, from 30h to 7Fh, are addressable as bytes.

Table 1.2.2 Internal RAM organization

Serial Number	Areas	Address	Function
1	Working registers area	0x00~0x07	Bank 0 (R0~R7)
		0x08~0x0F	Bank 1 (R0~R7)
		0x10~0x17	Bank 2 (R0~R7)
		0x18~0x1F	Bank 3 (R0~R7)
2	Bit-addressable area	0x20~0x2F	Bit-addressable area, Bit address: 0x00~0x7F
3	General-purpose RAM area	0x30~0x7F	User data buffer area

Special Function Registers

The 8051 operations that do not use the internal 128-byte RAM addresses from 00h to 7Fh are done by a group of specific internal registers, each called a special-function register (SFR), which may be addressed much like internal RAM, using addresses from 80h to FFh.

Some SFRs are also bit addressable, as is the case for the bit area of RAM. This feature allows the programmer to change only what needs to be altered, leaving the remaining bits in that SFR unchanged.

Not all of the addresses from 80h to FFh are used for SFRs, and attempting to use an address that is not defined or "empty", results in unpredictable results. The SFR names and equivalent internal RAM addresses are given in the Table 1.2.3:

Table 1.2.3   8051 SFR Address

Name	Function	Internal RAM Address (HEX)
A	Accumulator	E0
B	Arithmetic	F0
DPH	Addressing external memory	83
DPL	Addressing external memory	83
IE	Interrupt enable control	A8
IP	Interrupt priority	B8
P0	Input/output port latch	80
P1	Input/output port latch	90
P2	Input/output port latch	A0
P3	Input/output port latch	B0
PCON	Power control	87
PSW	Program status word	D0
SCON	Serial port control	98
SBUF	Serial port data buffer	99
SP	Stack pointer	81
TMOD	Timer/ counter mode control	89
TCON	Timer/ counter control	88
TL0	Timer 0 low byte	8A
TH0	Timer 0 high byte	8C
TL1	Timer 1 low byte	8B
TH1	Timer 1 high byte	8D

Note that the PC is not part of the SFR and has no internal RAM address.

SFRs are named in certain opcodes by their functional names such as A or TH0, and are referenced by other opcodes by their addresses, such as 0E0h or 8Ch. Note that any address used in the program must start with a number; thus address E0h for the A SFR begins with 0. Failure to use this number convention will result in an assembler error when the program is assembled.

Bit Definitions

The data type sbit is provided for the 8051 family and is used to declare an individual bit within the SFR of the 8051 family. For example, using this data type one can access to the individual bits of an I/O port.

Example:

sbit switch = P1^3;   /* variable switch is assigned to bit 3 of port 1 */
switch = 0;   /* clear bit 3 of port 1 */

The data type sfr is similar to sbit but is used to declare 8-bit variables.

Example:

sfr P1 = 0x90;   /* Port 1 address 0x90 assigned to P1 */
sfr P2 = 0xA0;   /* Port 2 address 0xA0 assigned to P2 */
unsigned char my_data;   /* declare my_data as unsigned character */
my_data = P1;   /* read 8 bit data from port 1 and assign to my_data */
P2 = my_data++;   /* increment my_data and send to port 2 */

Internal ROM

The 8051 is organized so that data memory and program code memory can be in two entirely different physical memory entities. Each has the same address ranges.

The structure of the internal RAM has been discussed previously. A corresponding block program code, contained in an internal ROM, occupies code address space 0000h to 0FFFh. The PC is ordinarily used to address program code bytes from addresses 0000h to FFFFh. Program addresses higher than 0FFFh, which exceed the internal ROM capacity, will cause the 8051 to automatically fetch code bytes from external program memory. Code bytes can also be fetched exclusively from an external memory, addresses 0000h to FFFFh, by connecting the external access pin (EA pin 31 on the DIP) to ground. The PC does not care where the code is; the circuit designer decides whether the code is found totally in internal ROM, totally in external ROM, or in a combination of internal and external ROM.

2.4 Structure of a Microcontroller-based C Program

The structure of a C program developed for a microcontroller is basically the same as the structure of a standard C program, with a few minor changes. The structure of a typical microcontroller-based C program is shown below. It is always advisable to describe the project at the beginning of a program using comment lines. The project name, filename, date, and the target processor type should also be included in this part of the program. The register definition file should then be included for the type of target processor used. This file is supplied as part of the compiler and includes the definitions for various registers of the microcontroller. In the example below, the register definition file for the Atmel 89C51 type microcontroller is included.

The global definitions of the variables used should then be entered, one line for each definition. The functions used in the program should then be included with the appropriate comments added to the heading and also to each line of the functions. The main program starts with the keyword main(), followed by the opening brackets '{'. The lines of the main program should also contain comments to clarify the operation of the program. The program is terminated by a closing bracket '}'.

```
/***********************************************************************
Project: Give project name
File: Give filename
Processor: Give target processor type
This is the program header. Describe your program here briefly.
***********************************************************************/
#include <reg51.h>

#define ......              /* include your define statements here */

sbit ......                 /* include your bit definitions here */

int ......
char ......                 /* include your global declarations here */

void func1()                /* include your functions here */
{
}

main()
{
                            /* main code */

                            /* include comments here */
}
```

2.5 The Basic of C Language: Data Types and Operators

The C programming language is a general-purpose high-level programming language that offers efficient and compact code and provides elements of structured programming. Many controlling and monitoring-based applications can be solved more efficiently with C than with any other programming language. C was originally available on mainframe computers, mini computers, and personal computers (PCs). The C programming language is now available on most microcontrollers and microprocessors.

In this textbook, the industry standard C51 optimizing C compiler is used throughout. C51 is available on Windows-based operating systems and the compiler implements the American National Standards Institute (ANSI) standard for the C language.

There are many other high-level language compilers available for microcontrollers, including PASCAL, BASIC, and other C compilers. Some of these compilers are freely available as shareware products and some can be obtained from the internet with little cost. Also, some companies supply free limited capability compilers, mainly for evaluation purposes. These compilers can be used for learning the features of a specific product and in some cases small projects can be developed with such compilers.

The C51 compiler has been developed for the 8051 family of microcontrollers. This is one of the most commonly used industry standard C compilers for the 8051 family, and can generate machine code for most of the 20-pin and 40-pin 8051 devices and its derivatives.

C51 is a professional, industry standard compiler with many features, including a large number of built-in functions. In this textbook, we shall be looking at the features of the C51 programming language as applied to programming single chip microcontrollers.

2.5.1 Data Types

Scan the QR code to view the teaching note of Data and Constants

Scan the QR code to watch the teaching video of Data and Constants

The C51 compiler provides the standard C data types and in addition several extended data types are offered to support the 8051 microcontroller family. Table 1.2.4 lists the available data types. Some of the data types are described below in more detail.

Table 1.2.4 CSI data types

Data type	Bits	Range
bit	1	0 or 1
signed char	8	−128 to +127
unsigned char	8	0 to +255
enum	16	−32,768 to +32,767
signed short	16	−32,768 to +32,767
unsigned short	16	0 to 65,535
signed int	16	−32,768 to +32,767
Unsigned int	16	0 to 65,535
signed long	32	−2,147,483,648 to 2,147,483,647
unsigned long	32	0 to 4,294,967,295
float	32	$\pm 1.175,494 \times 10^{-38}$ to $\pm 3.402,823 \times 10^{38}$
sbit	1	0 or 1
sfr	8	0 to 255
sfr16	16	0 to 65,535

Bit

These data types may be used to declare 1-bit variables.

Example:

bit my_flag;   /* declare my_flag as a bit variable */
my_flag 1;   /* set my_flag to 1 */

Signed Char/Unsigned Char

These data types are as in standard C language and are used to declare signed and unsigned character variables. Each character variable is 1 byte long (8 bits). Signed character variables range from −128 to +127, while unsigned character variables range from 0 to 255.

Example:

unsigned char var1, var2;   /* declare var1 and var2 as unsigned char */
var1 = 0xA4;   /* assign hexadecimal A4 to variable var1 */

var2 = var1;   /* assign var1 to var2 */

Signed Short/Unsigned Short

These data types are as in standard C language and are used to declare signed and unsigned short variables. Each short variable is 2 bytes long (16 bits). Signed short variables range from −32,768 to +32,767 and unsigned short variables range from 0 to 65,535.

Example:

unsigned short temp;   /* declare temp as unsigned short */
unsigned short wind;   /* declare wind as unsigned short */
temp = 0x0C200;   /* assign hexadecimal C200 to variable temp */
wind = temp;   /* assign variable temp to wind */

Signed Int/Unsigned Int

As in the standard C language, these data types are used to declare signed and unsigned integer variables. Integer variables are 2 bytes long (16 bits). Signed integers range from −32,768 to +32,767 and unsigned integers range from 0 to 65,535.

Example:

unsigned int n1, n2;   /* declare n1 and n2 as unsigned integers */
n1 = 10;   /* assign 10 to n1 */
n2 = 2*n1;   /* multiply n1 by 2 and assign to n2 */

Signed Long/Unsigned Long

These data types are as in standard C language and are used to declare signed and unsigned long integer variables. Each long integer variable is 4 bytes long (32 bits).

Example:

unsigned long temp;   /* declare temp as long integer variable */
temp = 250000;   /* assign 250000 to variable temp */

Float

This data type is used to declare a floating point variable.

Example:

float t1,t2;   /* declare t1 and t2 as floating point variables */
t1 = 25.4;   /* assign 25.4 to t1 */
t2 = sqrt(t1);   /* assign the square-root of t1 to t2 */

2.5.2 Constants

An **integer constant** like 1234 is an int. A long constant is written with a terminal l (ell) or L, as in 123,456,789L; an integer constant too big to fit into an int will also be taken as a long. Unsigned constants are written with a terminal u or U, and the suffix ul or UL indicates unsigned long.

Floating-point constants contain a decimal point (123.4) or an exponent (1e–2) or both; their type is double, unless suffixed. The suffixes f or F indicate a float constant; l or L indicate a long double.

The value of an integer can be specified in octal or hexadecimal instead of decimal. A leading 0 (zero) on an integer constant means octal; a leading 0x or 0X means hexadecimal. For example, decimal 31 can be written as 037 in octal and 0x1f or 0x1F in hex. Octal and hexadecimal constants may also be followed by L to make them long and U to make them unsigned: 0XFUL is an unsigned long constant with value 15 decimal.

A **character constant** is an integer, written as one character within single quotes, such as 'x'. The value of a character constant is the numeric value of the character in the machine's character set. For example, in the ASCII character set the character constant '0' has the value 48, which is unrelated to the numeric value 0. If we write '0' instead of a numeric value like 48 that depends on the character set, the program is independent of the particular value and easier to read. Character constants participate in numeric operations just as any other integers, although they are most often used in comparisons with other characters.

Certain characters can be represented in character and string constants by escape sequences like \n (newline); these sequences look like two characters, but represent only one. In addition, an arbitrary byte-sized bit pattern can be specified by

'\ooo'

where ooo is one to three octal digits (0...7) or by

'\xhh'

where hh is one or more hexadecimal digits (0...9, a...f, A...F). So we might write

#define VTAB '\013'   /* ASCII vertical tab */
#define BELL '\007'   /* ASCII bell character */

or, in hexadecimal,

#define VTAB '\xb'   /* ASCII vertical tab */
#define BELL '\x7'   /* ASCII bell character */

The complete set of escape sequences is shown in Table 1.2.5.

Table 1.2.5 Escape sequences

Escape sequences	Function	Escape sequences	Function
\a	Alert (bell) character	\\	Backslash
\b	Backspace	\?	Question mark
\f	Formfeed	\'	Single quote
\n	Newline	\"	Double quote
\r	Carriage return	\ooo	Octal number
\t	Horizontal tab	\xhh	Hexadecimal number
\v	Vertical tab		

The character constant '\0' represents the character with value zero, the null character. '\0' is often written instead of 0 to emphasize the character nature of some expression, but the numeric value is just 0.

A **constant expression** is an expression that involves only constants. Such expressions may be evaluated at compilation rather than run-time, and accordingly may be used in any place that a constant can occur, as in

#define MAXLINE 1000
char line[MAXLINE+1];

or

```
#define LEAP 1   /* in leap years */
int days[31+28+LEAP+31+30+31+30+31+31+30+31+30+31];
```

A **string constant**, or string literal, is a sequence of zero or more characters surrounded by double quotes, as in

"I am a string"

or

" "   /* the empty string */

The quotes are not part of the string, but serve only to delimit it. The same escape sequences used in character constants apply in strings; \" represents the double-quote character. String constants can be concatenated at compile time:

"hello," "world"

is equivalent to

"hello, world"

This is useful for splitting up long strings across several source lines.

Technically, a string constant is an array of characters. The internal representation of a string has a null character '\0' at the end, so the physical storage required is one more than the number of characters written between the quotes. This representation means that there is no limit to how long a string can be, but programs must scan a string completely to determine its length. The standard library function strlen(s) returns the length of its character string arguments, excluding the terminal '\0'. Here is our version:

```
/* strlen: return length of s */
int strlen(char s[ ])
{
    int i;
    while (s[i] != '\0')
```

```
    ++i;
    return i;
}
```

Strlen and other string functions are declared in the standard header <string.h>.

Be careful to distinguish between a character constant and a string that contains a single character: 'x' is not the same as "x". The former is an integer, used to produce the numeric value of the letter x in the machine's character set. The latter is an array of characters that contains one character (the letter x) and a '\0'.

There is one other kind of constant, the **enumeration constant**. An enumeration is a list of constant integer values, as in

enum boolean { NO, YES };

The first name in an enum has value 0, the next 1, and so on, unless explicit values are specified. If not all values are specified, unspecified values continue the progression from the last specified value, as the second of these examples:

```
enum escapes { BELL = '\a', BACKSPACE = '\b', TAB = '\t',
    NEWLINE = '\n', VTAB = '\v', RETURN = '\r' };
enum months { JAN = 1, FEB, MAR, APR, MAY, JUN,
              JUL, AUG, SEP, OCT, NOV, DEC };
              /* FEB = 2, MAR = 3, etc. */
```

Names in different enumerations must be distinct. Values need not be distinct in the same enumeration.

Enumerations provide a convenient way to associate constant values with names, an alternative to #define with the advantage that the values can be generated for you. Although variables of enum types may be declared, compilers need not check that what you store in such a variable is a valid value for the enumeration. Nevertheless, enumeration variables offer the chance of checking and so are often better than #defines. In addition, a debugger may be able to print values of enumeration variables in their symbolic form.

2.5.3 Declarations

Scan the QR code to view the teaching note of Declaration Operator and Expression   Scan the QR code to watch the teaching video of Declaration Operator and Expression

All variables must be declared before use, although certain declarations can be made implicitly by contents. A declaration specifies a type, and contains a list of one or more variables of that type, as in

    int lower, upper, step;
    char c, line[1000];

Variables can be distributed among declarations in any fashion; the lists above could well be written as

    int lower;
    int upper;
    int step;
    char c;
    char line[1000];

The latter form takes more space, but is convenient for adding a comment to each declaration for subsequent modifications.

A variable may also be initialized in its declaration. If the name is followed by an equals sign and an expression, the expression serves as an initializer, as in

    char esc = '\\';
    int i = 0;
    int limit = MAXLINE+1;
    float eps = 1.0e-5;

If the variable in question is not automatic, the initialization is done once only, conceptionally before the program starts executing, and the initializer must be a constant expression. An explicitly initialized automatic variable is initialized each time the function or block it is in is entered; the initializer may be any expression. External and static variables are initialized to zero by default. Automatic variables for which is no explicit initializer have undefined (i.e., garbage) values.

The qualifier const can be applied to the declaration of any variable to specify that its value will not be changed. For an array, the const qualifier says that the elements will not be altered.

const double e = 2.718,281,828,459,05;
const char msg[ ] = "warning:";

The const declaration can also be used with array arguments, to indicate that the function does not change that array:

int strlen(const char[ ]);

The result is implementation-defined if an attempt is made to change a const.

2.5.4 Operators
Arithmetic Operators

The binary arithmetic operators are +, –, *, /, and the modulus operator %. (See Table 1.2.6)

Table 1.2.6 Arithmetic operators

Operator	Meaning
+	Addition or unary plus
–	Subtraction or unary minus
*	Multiplication
/	Division
%	Modulo division

Integer division truncates any fractional part. The expression

x % y

produces the remainder when x is divided by y, and thus is zero when y divides x exactly. For example, a year is a leap year if it is divisible by 4 but not by 100, except that years divisible by 400 are leap years. Therefore

```
if ((year % 4 == 0 && year % 100 != 0) || year % 400 == 0)
    printf("%d is a leap year\n", year);
else
    printf("%d is not a leap year\n", year);
```

The % operator can not be applied to a float or double. The direction of truncation for / and the sign of the result for % are machine-dependent for negative operands, as is the action taken on overflow or underflow.

The binary + and − operators have the same precedence, which is lower than the precedence of *, / and %, which is in turn lower than unary + and −. Arithmetic operators associate left to right.

Relational and Logical Operators

The relational operators (Table 1.2.7) are

> >= < <=

They all have the same precedence. Just below them in precedence are the equality operators:

== !=

Relational operators have lower precedence than arithmetic operators, so an expression like i < lim−1 is taken as i < (lim−1), as would be expected.

More interesting are the logical operators (Table 1.2.8) && and ||. Expressions connected by && or || are evaluated left to right, and evaluation stops as soon as the truth or falsehood of the result is known. Most C programs rely on these properties. See Table 1.2.9 Truth table.

By definition, the numeric value of a relational or logical expression is 1 if the relation is true, and 0 if the relation is false.

The unary negation operator ! converts a non-zero operand into 0, and a zero operand in 1. A common use of ! is in constructions like if (!valid)

rather than

if (valid == 0)

It's hard to generalize about which form is better. Constructions like invalid read nicely ("if not valid"), but more complicated ones can be hard to understand.

Table 1.2.7 Relational operators

Operators	Meaning
<	is less than
<=	is less than or equal to
>	is greater than
>=	is greater than or equal to
==	is equal to
!=	is not equal to

Table 1.2.8 Logical operators

Operators	Meaning
&&	Logical AND
||	Logical OR
!	Logical NOT

Table 1.2.9 Truth table

a	b	!a	a&&b	a||b
1	1	0	1	1
1	0	0	0	1
0	1	1	0	1
0	0	1	0	0

Increment and Decrement Operators

C provides two unusual operators for incrementing and decrementing variables. The increment operator ++ adds 1 to its operand, while the decrement operator -- subtracts 1. We have frequently used ++ to increment variables, as in

if (c == '\n')

++n;

The unusual aspect is that ++ and -- may be used either as prefix operators (before the variable, as in ++n), or postfix operators (after the variable: n++). In both cases, the effect is to increment n. But the expression ++n increments n before its value is used, while n++ increments n after its value has been used. This means that in a context where the value is being used, not just the effect, ++n and n++ are different. If n is 5, then

x = n++;

sets x to 5, but

x = ++n;

sets x to 5. In both cases, n becomes 6. The increment and decrement operators can only be applied to variables. An expression like (i+j)++ is illegal.

Bitwise Operators

C provides six operators for bit manipulation (See Table 1.2.10). These may only be applied to integral operands, that is, char, short, int, and long, whether signed or unsigned.

Table 1.2.10   Bitwise operators

Operators	Meaning
&	Bitwise AND
|	Bitwise inclusive OR
^	Bitwise exclusive OR
<<	Left shift
>>	Right shift
~	One's complement (unary)

The bitwise AND operator & is often used to mask off some set of bits, for example

n = n & 0177;

sets to zero all but the low-order 7 bits of n.

The bitwise OR operator | is used to turn bits on:

x = x | SET_ON;

sets to one in x the bits that are set to one in SET_ON.

The bitwise exclusive OR operator ^ sets a one in each bit position where its operands have different bits, and zero where they are the same.

One must distinguish the bitwise operators & and | from the logical operators && and ||, which imply left-to-right evaluation of a truth value. For example, if x is 1 and y is 2, then x & y is zero while x && y is one.

The shift operators << and >> perform left and right shifts of their left operand by the number of bit positions given by the right operand, which must be non-negative. Thus x << 2 shifts the value of x by two positions, filling vacated bits with zero. This is equivalent to multiplication by 4. Right shifting an unsigned quantity always fits the vacated bits with zero. Right shifting a signed quantity will fill with bit signs ("arithmetic shift") on some machines and with 0-bits ("logical shift") on others.

The unary operator ~ yields the one's complement of an integer. That is, it converts each 1-bit into a 0-bit and vice versa. For example

x = x & ~077

sets the last six bits of x to zero. Note that x & ~077 is independent of word length, and is thus preferable to, for example, x & 0 177,700, which assumes that x is a 16-bit quantity. The portable form involves no extra cost, since ~077 is a constant expression that can be evaluated at compile time.

Assignment Operators and Expressions

An expression such as

i = i + 2

in which the variable on the left side is repeated immediately on the right, can be written in the compressed form

i += 2

The operator += is called an assignment operator.

Most binary operators (operators like + that have a left and right operand) have a corresponding assignment operator op=, where op is one of

+   -   *   /   %   <<   >>   &   ^ |

If expr1 and expr2 are expressions, then

expr1 op= expr2

is equivalent to

expr1 = (expr1) op (expr2)

except that expr1 is computed only once. Note that the parentheses around expr2:

x *= y + 1

means

x = x * (y + 1)

rather than

x = x * y + 1

2.6 The Basic of C Language: Iteration Statements—for and while

Scan the QR code to view the teaching note of Iteration Statement

Scan the QR code to watch the teaching video of Iteration Statement

The control-flow of a language specifies the order in which computations are performed.

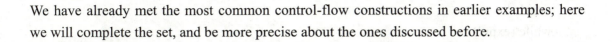

We have already met the most common control-flow constructions in earlier examples; here we will complete the set, and be more precise about the ones discussed before.

2.6.1 Statements and Blocks

An expression such as x = 0 or i++ or printf(...) becomes a statement when it is followed by a semicolon, as in

x = 0;
i++;
printf(...);

In C, the semicolon is a statement terminator, rather than a separator as it is in languages like Pascal.

Braces { and } are used to group declarations and statements together into a compound statement, or block, so that they are syntactically equivalent to a single statement. The braces that surround the statements of a function are one obvious example; braces around multiple statements after an if, else, while, or for are another. (Variables can be declared inside any block) There is no semicolon after the right brace that ends a block.

2.6.2 Loops—while and for

We have already encountered the while and for loops. In

while (expression)
statement

the expression is evaluated. If it is non-zero, statement is executed and expression is reevaluated. This cycle continues until expression becomes zero, at which point execution resumes after statement.

The **for** statement

for (expr1; expr2; expr3)
statement

is equivalent to

```
expr1;
while (expr2)
{
    statement
    expr3;
}
```

except for the behaviour of continue.

Grammatically, the three components of a for loop are expressions. Most commonly, expr1 and expr3 are assignments or function calls and expr2 is a relational expression. Any of the three parts can be omitted, although the semicolons must remain. If expr1 or expr3 is omitted, it is simply dropped from the expansion. If the test, expr2, is not present, it is taken as permanently true, so

```
for (;;)
{
    ...
}
```

is an "infinite" loop, presumably to be broken by other means, such as a break or return.

Whether to use while or for is largely a matter of personal preference. For example, in

```
while ((c = getchar()) == ' ' || c == '\n' || c = '\t')
    ;       /* skip white space characters */
```

there is no initialization or re-initialization, so the while is most natural.

The for is preferable when there is a simple initialization and increment since it keeps the loop control statements close together and visible at the top of the loop. This is most obvious in

```
for (i = 0; i < n; i++)
    ...
```

which is the C idiom for processing the first n elements of an array.

2.6.3 Loops—do-while

The while and for loops test the termination condition at the top. By contrast, the third loop in C, the do-while, tests at the bottom after making each pass through the loop body; the body is always executed at least once.

The syntax of the do is

do
   statement
while (expression);

The statement is executed, then expression is evaluated. If it is true, statement is evaluated again, and so on. When the expression becomes false, the loop terminates. Experience shows that do-while is much less used than while and for. Nonetheless, from time to time it is valuable.

2.6.4 Break and Continue

It is sometimes convenient to be able to exit from a loop other than by testing at the top or bottom. The break statement provides an early exit from for, while, and do, just as from switch. A break causes the innermost enclosing loop or switch to be exited immediately.

The following function, trim, removes trailing blanks, tabs and newlines from the end of a string, using a break to exit from a loop when the rightmost non-blank, non-tab, non-newline is found.

```
/* trim: remove trailing blanks, tabs, newlines */
int trim(char s[])
{
    int n;
    for (n = strlen(s)–1; n >= 0; n--)
        if (s[n] != ' ' && s[n] != '\t' && s[n] != '\n')
            break;
    s[n+1] = '\0';
    return n;
}
```

Strlen returns the length of the string. The for loop starts at the end and scans backwards looking for the first character that is not a blank or tab or newline. The loop is broken when one is found, or when n becomes negative (that is, when the entire string has been scanned). You should verify that this is correct behavior even when the string is empty or contains only white space characters.

The continue statement is related to break, but less often used; it causes the next iteration of the enclosing for, while, or do loop to begin. In the while and do, this means that the test part is executed immediately; in the for, control passes to the increment step. The continue statement applies only to loops, not to switch. A continue inside a switch inside a loop causes the next loop iteration.

For example, this fragment processes only the non-negative elements in the array; negative values are skipped.

```
for (i = 0; i < n; i++)
    if (a[i] < 0)    /* skip negative elements */
        continue;
    ...              /* do positive elements */
```

The continue statement is often used when the part of the loop that follows is complicated, so that reversing a test and indenting another level would nest the program too deeply.

2.7 The Basic of C Language: Functions

Behind all well-written programs in the C programming language lies the same fundamental element—the function. You've used functions in every program that you've encountered so far. Indeed, each and every program also used a function called main. So you might ask, what is all the fuss about. The truth is that the program function provides the mechanism for producing programs that are easy to write, read, understand, debug, modify, and maintain. Obviously, anything that can accomplish all of these things is worthy of a bit of fanfare.

2.7.1 Defining a Function

First, you must understand what a function is, and then you can proceed to find out how it can be the most effectively used in the development of programs.

```
#include <reg51.h>
void main (void)
```

```
{
    int i;
    P0 = 0xff;
    for(i=0; i<1000; i++);
    P0 = 0x00;
    for(i=0; i<1000; i++);
}
```

Here is a function called delay that does the same thing:

```
#include <reg51.h>
void delay (void)
{
    int i;
    for(i=0; i<1000; i++);
}
void main (void)
{
    int i;
    P0 = 0xff;
    delay();
    P0 = 0x00;
    delay();
}
```

The first line of a function definition tells the compiler (in order from left to right) four things about the function:

- Who can call it;
- The type of value it returns;
- Its name;
- The arguments it takes.

The first line of the delay function definition tells the compiler that the function returns no value (the first use of the keyword void), its name is delay, and that it takes no arguments (the second use of the keyword void). You learn more details about the void keyword shortly.

Obviously, choosing meaningful function names is just as important as choosing meaningful variable names—the choice of names greatly affects the program's readability.

Program above consists of two functions: delay and main. Program execution always begins with main. Inside that function, the statement

delay ();

appears. This statement indicates that the function delay is to be executed. The open and close parentheses are used to tell the compiler that delay is a function and that no arguments or values are to be passed to this function (which is consistent with the way the function is defined in the program). When a function call is executed, program execution is transferred directly to the indicated function. After the for loop has been finished, the delay routine is finished too (as signaled by the closing brace) and the program returns to the main routine, where program execution continues at the point where the function call was executed. Note that it is acceptable to insert a return statement at the end of delay like this:

return;

Because delay does not return a value, no value is specified for the return. This statement is optional because reaching the end of a function without executing a return has the effect of exiting the function anyway without returning a value. In other words, either with or without the return statement, the behavior on exit from delay is identical.

2.7.2 Arguments and Local Variables

```
void delay (unsigned int i)
{
    unsigned int k;
    for(k=0; k<i; k++);
}
```

The function above contains an argument i, we can change the delay time by setting it. Arguments, greatly increase the usefulness and flexibility of a function.

2.7.3 Function Prototype Declaration

The function delay requires a bit of explanation. The first line of the function:

void delay (unsigned int i)

is called the function prototype declaration. It tells the compiler that delay is a function that returns no value (the keyword void) and that takes a single argument, called i, which is an unsigned int. The name that is chosen for an argument, called its formal parameter name, as well as the name of the function itself, can be any valid name formed by observing the rules outlined for forming variable names. For obvious reasons, you should choose meaningful names.

After the formal parameter name has been defined, it can be used to refer to the argument anywhere inside the body of the function. The beginning of the function's definition is indicated by the opening curly brace.

The beginning of the function's definition is indicated by the opening curly brace. You need a variable to act as your loop index. The variable k is defined for the purpose and are declared to be of type unsigned int. This variable is defined and initialized in the same manner that you defined and initialized your variables inside the main routine in previous programs.

Circuit Diagram

As is shown in Figure 1.2.8, the circuit consists of the MCU and eight LEDs connected to Port 2 of the microcontroller.

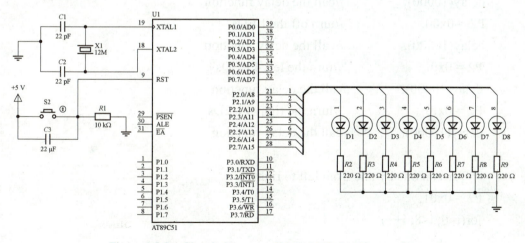

Figure 1.2.8　Circuit diagram of eight LEDs basic operation

Program Description

/***

PROJECT: Eight LEDs Basic Operation

FILE: PROJ1_2.C

MCU Practical Technology

PROCESSOR: AT89C51
This project controls eight LEDs turn on and turn off.
**/
#include <reg51.h>

/*Function to delay*/
void delay(unsigned int i)
{
    unsigned int k;
    for(k=0;k<i;k++);
}

/* start of main program */
void main()
{
    char i;   //declare a variable
    while(1)
    {
        P2 = 0xff;   //turn on the LEDs
        delay(10000);   //call the delay function
        P2 = 0x00;   //turn off the LEDs
        delay(10000);   //call the delay function
        P2 = 0x0f;   //turn the left 4 LEDs
        delay(10000);   //call the delay function
        P2 = 0xf0;   //turn the right 4 LEDs
        delay(10000);   //call the delay function

        /* Turn on one by one from left to right */
        P2 = 0x01;
        for(i=0; i<8; i++)
        {
            P2 = P2 << 1;
            delay(10000);
        }
    }
}

Project 1　Household Lighting Control System Design

Quiz

1. Once the MCU is in a chaotic state, which signal is effective to make it instantly from chaos back to the original harmony? (　　)
   A) Clock signal.
   B) Interrupt the system.
   C) The timer.
   D) Reset signal.

2. What is the condition of MCU reset? (　　)
   A) There is a high level above 1 machine cycle on the reset pin.
   B) There is a high level on the reset pin with more than 2 machine cycles.
   C) There is a low level above 1 machine cycle on the reset pin.
   D) Reset the low level on the pin with more than 2 machine cycles.

3. Which of the following is the reset pin of MCU? (　　)
   A) RST.
   B) XTAL1.
   C) XTAL2.
   D) RXD.

4. What are the three program structures of C?(　　)
   A) Sequential structure.
   B) Selective structure.
   C) Loop structure.

5. Programs are stored in program memory in form (　　).
   A) C language source program
   B) assembler
   C) binary coding
   D) BCD

Scan the QR code to view the answer of quiz

Summary

This project introduces the C51 program structure, basic statements, data types, operators and expressions and functions, through a series of project tasks, training C51 structured program design method and MCU application system design method. Key contents to be mastered in this project include:

- C language is a structured programming language, which has three basic program structures: sequential structure, selection structure and loop structure. Besides, it has abundant operators and data types oriented to the hardware structure of single chip microcomputer, so it has strong processing ability.
- The basic statements of C language include expression statement, assignment statement, if statement, switch statement, while statement, do-while statement and for statement.
- In addition to all the standard data types of ANSI C, C51 extends some special data types: bit, sbit, sfr and sfr16, which are used to access the special registers and addressable locations of the single-chip microcomputer, in order to make more effective use of the hardware resources of 51 MCU.
- Function is the basic unit of C statement program, a C source program includes at least one function, a C source program has and only one main function main(). C programs always start with the main() function.
- C51 functions are divided into internal standard functions and user-defined functions. Any function, like a variable, must be defined before it can be used. Calling an internal function requires the C program to include the header file name of the function prototype with the prepossessing command "#include". Custom functions can be defined before a function is called, or declared and used first.

Task 3   Button Interface

Objectives

After completing this task, you should be able to
- through the button control LED display of different patterns, understand the button interface, and the key control program design method.
- configure I/O pins for different operation modes.
- interface with keyboard.

Project 1　Household Lighting Control System Design

Task Requirement

One button S1 controls 8 LEDs to display different patterns. When S1 is not pressed, 8 LEDs are fully turned on; when S1 is pressed, 8 LEDs cross on and off.

3.1　MCU Parallel I/O Port

Scan the QR code to view the teaching note of MCU Parallel I/O Port

Scan the QR code to watch the teaching video of MCU Parallel I/O Port

Input/Output Pins Ports, and Circuits

One major feature of a microcontroller is the versatility built into the input/output (I/O) circuits that connect the 8051 to the outside world. Microprocessor designs must add additional chips to interface with external circuitry. This ability is built into the microcontroller.

To be commercially viable, the 8051 has to incorporate as many functions as were technically and economically feasible. The main constraint that limits numerous functions is the number of pins available to the 8051 circuit designers. The dip has 40 pins, and the success of the design in the marketplace was determined by the flexibility built into the use of these pins.

For this reason, 24 of the pins may each be used for one of the two entirely different functions, yielding a total pin configuration of 64. The function a pin performs at an given instant depends, first, upon what is physically connected to it and then, upon what software commands are used to program the pin. Both of these factors are under the complete control of the 8051 programmer and circuit designer.

Given this pin flexibility, the 8051 may be applied simply as a single component with I/O only, or it may be expanded to include additional memories, parallel ports, and serial data communication by using the alternate pin assignments. The key to programming an alternate pin function is the port pin circuitry shown in Figure 1.3.1.

Each port has a D-type output latch for each pin. The SFR for each port is made up of these eight latches which can be addressed at the sfr address for that port. For instance, the eight

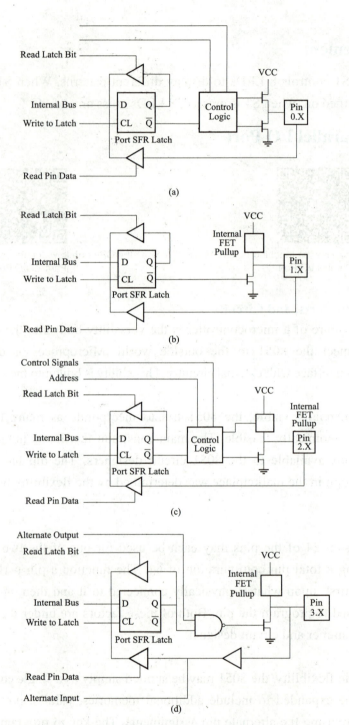

Figure 1.3.1   Port pin circuits

(a) Port 0 pin configuration; (b) Port 1 pin configuration; (c) Port 2 pin configuration; (d) Port 3 pin configuration

latches for port 0 are addressed at location 80h; port 0 pin 3 is bit 2 of the P0 SFR. The port latches should not be confused with the port pins; the data on the latches does not have to be the same as that on the pins.

Project 1　Household Lighting Control System Design

The two data paths are shown in Figure 1.3.1 by the circuits that read the latch or pin data using two entirely separate buffers. The top buffer is enabled when latch data is read and the lower buffer is enabled when the pin state is read. The status of each latch may be read from a latch buffer, while an input buffer is connected directly to each pin so that the pin status may be read independently of the latch state.

Different opcodes access the latch or pin states as appropriate. Port operations are determined by the manner in which the 8051 is connected to external circuitry.

Programmable port pins have completely different alternate functions. The configuration of the control circuitry between the output latch and the port pin determines the nature of any particular port pin functions. An inspection of Figure 1.3.1 reveals that only port 1 can not have alternate functions, while ports 0, 2, and 3 can be programmed.

Port 0

Port 0 pins may serve as inputs, outputs, or when used together, as a bi-directional low-order address and data bus for external memory. For example, when a pin is to be used as an input, a 1 must be written to the corresponding port 0 latch by the program thus turning both of the output transistors off, which in turn causes the pin to "float" in a high impedance state, and the pin is essentially connected to the input buffer.

When used as an output, the pin latches that are programmed to a 0 will turn on the lower FET, grounding the pin. All latches that are programmed to a 1 still float; thus external pull-up resistors will be needed to supply a logic high when using port 0 as an output.

When port 0 is used as an address bus to external memory, internal control signals switch the address lines to the gates of the Field Effect Transistors (FETs). A logic 1 on an address bit will turn the upper FET on and the lower FET off to provide a logic high at the pin. When the address bit is a zero, the lower FET is on and the upper FET off to provide a logic low at the pin. After the address has been formed and latched into external circuits by the Address latch Enable (ALE) pulse, the bus is turned around to become a data bus. Port 0 now reads data from the external memory and must be configured as an input, so a logic 1 is automatically written by internal control logic to all port 0 latches.

Port 1

Port 1 pins have no dual functions, therefore the output latch is connected directly to the gate of the lower FET, which has an FET circuit labeled "Internal FET Pullup" as an active pull-up load.

Used as an input, a 1 is written to the latch, turning the lower FET off; the pin and the input to the pin buffer are pulled high by the FET load. An external circuit can overcome the high impedance pull-up and drive the pin low to input a 0 or leave the input high for a 1.

If used as an output, the latches containing a 1 can drive the input of an external circuit high through the pull-up. If a 0 is written to the latch, the lower FET is on, the pull-up is off, and the pin can drive the input of the external circuit low.

To aid in speeding up switching times when the pin is used as an output, the internal FET pull-up has another FET in parallel with it. The second FET is turned on for two oscillator time periods during a low-to-high transition on the pin, as is shown in Figure 1.3.1(b). This arrangement provides a low impedance path to the positive voltage supply to help reduce rise times in charging any parasitic capacitances in the external circuitry.

Port 2

Port 2 may be used as an input/output port in operation similar to port 1. The alternate use of port 2 is to supply a high-order address byte in conjunction with the port 0 low-order byte to address external memory.

Port 2 pins are momentarily changed by the address control signals when supplying the high byte of a 16-bit address. Port 2 latches remain stable when external memory is addressed, as they do not have to be turned around (set to 1) for data input, as is the case for port 0.

Port 3

Port 3 is an input/output port similar to port 1. The input and output functions can be programmed under the control of the P3 latches or under the control of various other special function registers. The port 3 alternate uses are shown in the Table 1.3.1.

Table 1.3.1 Port 3 alternate uses

PIN	ALTERNATE USE	SFR
P3.0/RXD	Serial data input	SBUF
P3.1/TXD	Serial data output	SBUF
P3.2/$\overline{INT0}$	External interrupt 0	TCON.1
P3.3/$\overline{INT1}$	External interrupt 1	TCON.3
P3.4/T0	External timer 0 input	TMOD
P3.5/T1	External timer 1 input	TMOD
P3.6/$\overline{WR}$	External memory write pulse	—
P3.7/$\overline{RD}$	External memory read pulse	—

Unlike ports 0 and 2, which can have external addressing functions and change all eight port bits when in alternate use, each pin of port 3 may be individually programmed to be used either as I/O or as one of the alternate functions.

3.2 Use of Individual Buttons

Scan the QR code to view the teaching note of Use of Individual Buttons

Scan the QR code to watch the teaching video of Use of Individual Buttons

In the brief moment when the key is pressed, the jitter on the hardware will often produce a jitter of several milliseconds. If the signal is collected at this time, it will inevitably lead to misoperation and even system collapse. Similarly, when the key is released, the hardware will shake accordingly, which will have the same consequences. Therefore, in analog or digital circuits, we should avoid collecting signals and operating at the most unstable time.

The ways to eliminate key jitter are listed as following:

- Software delay;
- Several low level count;
- Low pass filtering.

In the C51, we often use the first way to solve the problem. The process of software delay eliminating jitter is as follows (See Figure 1.3.2):

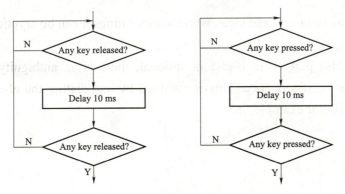

Figure 1.3.2 Key detection flow chart

- Signal detected;
- Delay 5–10 ms to eliminate jitter;
- Continue to detect the signal to confirm whether it is pressed;
- If yes, start waiting for release;
- If no, signal detected again.

3.3 The Basic of C language: Selection Statement—if-else

If-else

The if-else statement is used to express decisions. Formally the syntax is

if (expression)
    statement1
else
    statement2

where the else part is optional. The expression is evaluated; if it is true (that is, if expression has a non-zero value), statement1 is executed. If it is false (expression is zero) and if there is an else part, statement2 is executed instead.

Since an if tests the numeric value of an expression, certain coding shortcuts are possible. The most obvious is writing

    if (expression)

instead of

    if (expression != 0)

Sometimes this is natural and clear, while at other times it can be cryptic.

Because the else part of an if-else is optional, there is an ambiguity when an else if omitted from a nested if sequence. This is resolved by associating the else with the closest previous else-less if. For example, in

    if (n > 0)
        if (a > b)
            z = a;
        else

z = b;

the else goes to the inner if, as we have shown by indentation. If that isn't what you want, braces must be used to force the proper association:

```
if (n > 0)
{
    if (a > b)
        z = a;
}
else
    z = b;
```

The ambiguity is especially pernicious in situations like this:

```
if (n > 0)
    for (i = 0; i < n; i++)
        if (s[i] > 0)
        {
            printf("...");
            return i;
        }
    else        /* WRONG */
        printf("error -- n is negative\n");
```

The indentation shows unequivocally what you want, but the compiler doesn't get the message, and associates the else with the inner if. This kind of bug can be hard to find; it's a good idea to use braces when there are nested ifs.

By the way, notice that there is a semicolon after z = a in

```
if (a > b)
    z = a;
else
    z = b;
```

This is because grammatically, a statement follows the if, and an expression statement like "z = a;" is always terminated by a semicolon.

Else if

The construction

    if (expression)
        statement
    else if (expression)
        statement
    else if (expression)
        statement
    else if (expression)
        statement
    else
        statement

occurs so often that it is worth a brief separate discussion. This sequence of if statements is the most general way of writing a multi-way decision. The expressions are evaluated in order; if an expression is true, the statement associated with it is executed, and this terminates the whole chain. As always, the code for each statement is either a single statement, or a group of them in braces.

The last else part handles the "none of the above" or default case where none of the other conditions is satisfied. Sometimes there is no explicit action for the default; in that case the trailing

    else
        statement

can be omitted, or it may be used for error checking to catch an impossible condition.

```
/* Program to evaluate simple expressions of the form value operator value */
#include <stdio.h>
int main (void)
{
    float value1, value2;
    char operator;
```

```
        printf ("Type in your expression.\n");
        scanf ("%f %c %f", &value1, &operator, &value2);

        if ( operator == '+' )
            printf ("%.2f\n", value1 + value2);
        else if ( operator == '-' )
            printf ("%.2f\n", value1 - value2);
        else if ( operator == '*' )
            printf ("%.2f\n", value1 * value2);
        else if ( operator == '/' )
        {
            if ( value2 == 0 )
                printf ("Division by zero.\n");
            else
                printf ("%.2f\n", value1 / value2);
        }
        else
            printf ("Unknown operator.\n");

        return 0;
}
```

When the operator is typed in is the slash, for division, another test is made to determine if value2 is 0. If it is, an appropriate message is displayed at the terminal. Otherwise, the division operation is carried out and the results are displayed. Pay careful attention to the nesting of the if statements and the associated else clauses in this case.

The else clause at the end of the program catches any "fall throughs". Therefore, any value of the operator that does not match any of the four characters tested causes this else clause to be executed, resulting in the display of "Unknown operator".

Circuit Diagram

According to the function, 8 LEDs are connected to port 0, and the button is connected to P3.0. The circuit diagram of button interface is shown in the Figure 1.3.3.

P3.0 is connected with +5V power supply through pull-up resistance R_{10}. When S1 is not pressed, P3.0 remains high, while S1 is pressed, P3.0 is grounded. Therefore, you can tell if button S1 is pressed or not, through reading P3.0 state.

When 51 MCU is reset, I/O pins are in high level, so you can ignore the pull-up resistance, directly put the button S1 grounded.

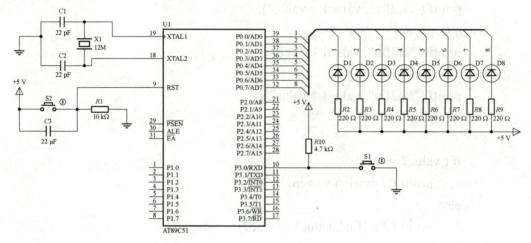

Figure 1.3.3   Circuit diagram of button interface

Program Description

```
/*********************************************************************
PROJECT: Button Interface
FILE: PROJ1_3.C
PROCESSOR: AT89C51
This project uses a button to control LEDs.
When S1 is not pressed, 8 LEDs are fully turned on;
when S1 is pressed, 8 LEDs cross on and off.
*********************************************************************/
#include <reg51.h>
sbit S1 = P3^0;

/*Function to delay*/
void delay(unsigned int i)
{
    unsigned int k;
    for(k=0;k<i;k++);
}
```

```
/* start of main program */
main()
{
    if(S1 == 0)              //If S1 is pressed
    {
        delay(10000);        //delay to debunce
        if(S1 == 0)          //If S1 is pressed
        {
            P0 = 0x55;       //8 LEDs cross on and off
        }
    }
    else
    {
        P0 = 0x00;           //8 LEDs are fully turned on
    }
}
```

Quiz

1. The name of the pin of the accessor store control signal is (   ).

    A) EA

    B) PSEN

    C) RST

    D) ALE

2. MCU parallel I/O port line P1.0 to P1.3 are connected to four LEDs, at this time it is (   ) interface function.

    A) Input

    B) Output

3. In 8051 MCU, the pin used in the input/output pin for the special second function is (   ).

    A) P0

    B) P1

    C) P2

    D) P3

4. When is the value of an expression in if false? (   )

A) 0.
B) 1.

5. If a key is defined as K, low when pressed, and high when released, (   ) of the following options describes the statement waiting for the key to be released.
   A) while (K);
   B) while (! K);
   C) while (K = = 0);
   D) while (K = = 1).

Scan the QR code to view the answer of quiz

Summary

The internal hardware configuration of the 8051 registers and control circuits have been examined at the functional block diagram level. The 8051 may be considered to be a collection of RAM, ROM, and addressable registers that have some unique functions.

Extended Reading

The development of Semiconductor in China

Domestic semiconductor development can be roughly divided into three stages.

1. The first stage is 1982—2000, which is called the frame stage.

In 1982, Huajing established the leading group for computer and large-scale integrated circuits. Due to the favorable international environment at that time, It proposed to exchange market for technology, and set up a semiconductor industry base with Beijing, Shanghai and Wuxi as the center, especially Huajing Wuxi in 1990s. Since then, Huajing has become a benchmark semiconductor enterprise attracting domestic attention.

2. The second stage is the initial stage of commercialization, 2000—2014.

The government of China provides preferential taxes and fiscal policies in semiconductor industry, to realize the industrial division.

After 2000, Tianjin MOTOROLA invested $1.4 billion to build an 8-inch plant producing 25,000 wafers a month, and Shanghai SMIC invested $1.5 billion to build an 8-inch plant producing 42,000 wafers a month.

By 2003, there were a number of wafer OEM enterprises in China, such as Shanghai Hongli, Suzhou Hejian (United Power), Shanghai Beiling, Shanghai Advanced (Philips), Beijing SMIC Global and so on.

3. The third stage is from 2014 to 2030, marking the beginning of a period of leapfrog development.

In June 2014, the State Council promulgated the Outline for Promoting the Development of the National Integrated Circuit Industry, proposing the establishment of the National Integrated Circuit Industry Fund (referred to as the "Big Fund") to elevate the research and development of new technologies in the semiconductor industry to a national strategic height.

In addition, it clearly states that by 2020, the gap between the integrated circuit industry and the international advanced level will be gradually narrowed, sales revenue of the whole industry will grow at an average annual rate of more than 20%, and the sustainable development ability of enterprises will be greatly enhanced.

By 2030, major links in the IC industry chain will reach the international advanced level, and a number of enterprises will enter the international first echelon, realizing leapfrog development.

Project 2

Household Access Control System Design

Task 1 Independent Button Access Control System Design

Objectives

After completing this task, you should be able to
- interface with 7-segment displays;
- use arrays and pointers for data manipulation;
- use 7-segment digital tube to display messages.

Task Requirement

This task designs a simple access control system including 4 buttons and 1 7-segment display driver.

In some intelligent access control systems, the correct password is required to unlock the lock. The hardware circuit consists of three parts: button input, digital display and electronic lock unlocking drive circuit.

The hardware circuit design: 4 buttons, controlled by P1.0–P1.3 of P1 port; a digital tube, static control by P0; one LED, controlled by the P3.0, LED on and off respectively to simulate the opening and locking of the unlock circuit.

The basic functions are as follows.

- 4 buttons, respectively representing the numbers 0,1,2,3.

- The password is set in advance in the program, with the number between 0 and 3.
- The 7-segment tube displays "?", indicating waiting for password input.
- When the password is entered correctly, the character "P" is displayed about 3s, and the lock is opened through P3.0; Otherwise, the character "E" is displayed about 3s, and the lock remains locked.

When the program is designed, set the initial password lock to close and the display symbol is "?". When the number key is pressed, if it is the same as the pre-set password, it will display the character "P". Otherwise, the character "E" display continues for 3s, holding the lock and waiting for the next password input. Use common anode LED digital tube static display mode, password set as "2".

1.1 7-segment Display Driver

Scan the QR code to view the teaching note of 7-segment Digital Tube Structure and Static Display

Scan the QR code to watch the teaching video of 7-segment Digital Tube Structure and Static Display

An LED or light emitting diode, is a solid state optical pn-junction diode which emits light energy in the form of photons. The emission of these photons occurs when the diode junction is forward biased by an external voltage allowing current to flow across its junction, and in electronics we call this process electroluminescence.

The actual colour of the visible light emitted by an LED, ranging from blue to red to orange, is decided by the spectral wavelength of the emitted light which itself is dependent upon the mixture of the various impurities added to the semiconductor materials used to produce it.

Light emitting diodes have many advantages over traditional bulbs and lamps, with the main ones being their small size, long life, various colours, cheapness and are readily available, as well as being easy to interface with various other electronic components and digital circuits.

But the main advantage of light emitting diodes is that because of their small size, several of them can be connected together within one small and compact package producing what is generally called a 7-segment display which is shown in Figure 2.1.1.

MCU Practical Technology

Figure 2.1.1 7-segment display

The 7-segment display, also written as "seven segment display", consists of seven LEDs (hence its name) arranged in a rectangular fashion, as is shown in Figure 2.1.1. Each of the seven LEDs is called a segment because when illuminated the segment forms part of a numerical digit (both Decimal and Hex) to be displayed. An additional 8th LED is sometimes used within the same package thus allowing the indication of a decimal point, (DP) when two or more 7-segment displays are connected together to display numbers greater than ten.

Each one of the seven LEDs in the display is given a positional segment with one of its connection pins being brought straight out of the rectangular plastic package. These individually LED pins are labelled from a to g representing each individual LED. The other LED pins are connected together and wired to form a common pin.

So by forward biasing the appropriate pins of the LED segments in a particular order, some segments will be light and others will be dark allowing the desired character pattern of the number to be generated on the display. This then allows us to display each of the ten decimal digits 0 to 9 on the same 7-segment display.

The displays common pin is generally used to identify which type of 7-segment display it is. As each LED has two connecting pins, one is called the "anode" and the other one is called the "Cathode", there are therefore two types of LED 7-segment display called common cathode (CC) and common anode (CA).

The difference between the two displays, as their name suggests, is that the common cathode has all the cathodes of the 7-segments connected directly together and the common anode has all the anodes of the 7-segments connected together and is illuminated as follows.

The Common Cathode (CC)

In the common cathode display, all the cathode connections of the LED segments are joined together to logic "0" or ground. The individual segments are illuminated by application of a "HIGH", or logic "1" signal via a current limiting resistor to forward bias the individual anode terminals (a–g). (See Figure 2.1.2)

Project 2 Household Access Control System Design

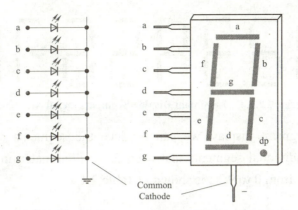

Figure 2.1.2 Common cathode 7-segment display

The Common Anode (CA)

In the common anode display, all the anode connections of the LED segments are joined together to logic "1". The individual segments are illuminated by applying a ground, logic "0" or "LOW" signal via a suitable current limiting resistor to the cathode of the particular segment (a–g). (See Figure 2.1.3)

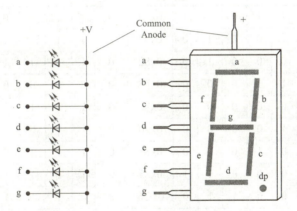

Figure 2.1.3 Common anode 7-segment display

In general, common anode displays are more popular as many logic circuits can sink more current than they can source. Also note that a common cathode display is not a direct replacement in a circuit for a common anode display and vice versa, as it is the same as connecting the LEDs in reverse, and hence light emission will not take place.

Depending upon the decimal digit to be displayed, the particular set of LEDs is forward biased. For instance, to display the numerical digit 0, we will need to light up six of the LED segments corresponding to a, b, c, d, e and f. Thus the various digits from 0 to 9 can be displayed using a 7-segment display, as is shown in Figure 2.1.4.

73

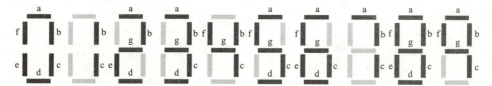

Figure 2.1.4 7-Segment Display Segments for all Numbers

Then for a 7-segment display, we can produce the segments and corresponding bit patterns giving the individual segments that need to be illuminated in order to produce the required decimal digit from 0 to 9, as is shown in Table 2.1.1.

Table 2.1.1 Segments and corresponding bit patterns

Number	The Common Anode Display								The Common Cathode Display							
	dp	g	f	e	d	c	b	a	hex	dp	g	f	e	d	c	b	a	hex
0	1	1	0	0	0	0	0	0	0xC0	0	0	1	1	1	1	1	1	0x3F
1	1	1	1	1	1	0	0	1	0xF9	0	0	0	0	0	1	1	0	0x06
2	1	0	1	0	0	1	0	0	0xA4	0	1	0	1	1	0	1	1	0x5B
3	1	0	1	1	0	0	0	0	0xB0	0	1	0	0	1	1	1	1	0x4F
4	1	0	0	1	1	0	0	1	0x99	0	1	1	0	0	1	1	0	0x66
5	1	0	0	1	0	0	1	0	0x92	0	1	1	0	1	1	0	1	0x6D
6	1	0	0	0	0	0	1	0	0x82	0	1	1	1	1	1	0	1	0x7D
7	1	1	1	1	1	0	0	0	0xF8	0	0	0	0	0	1	1	1	0x07
8	1	0	0	0	0	0	0	0	0x80	0	1	1	1	1	1	1	1	0x7F
9	1	0	0	1	0	0	0	0	0x90	0	1	1	0	1	1	1	1	0x6F
A	1	0	0	0	1	0	0	0	0x88	0	1	1	1	0	1	1	1	0x77
B	1	0	0	0	0	0	1	1	0x83	0	1	1	1	1	1	0	0	0x7C
C	1	1	0	0	0	1	1	0	0xC6	0	0	1	1	1	0	0	1	0x39
D	1	0	1	0	0	0	0	1	0xA1	0	1	0	1	1	1	1	0	0x5E
E	1	0	0	0	0	1	1	0	0x86	0	1	1	1	1	0	0	1	0x79
F	1	0	0	0	1	1	1	0	0x8E	0	1	1	1	0	0	0	1	0x71
H	1	0	0	0	1	0	0	1	0x89	0	1	1	1	0	1	1	0	0x76
L	1	1	0	0	0	1	1	1	0xC7	0	0	1	1	1	0	0	0	0x38
P	1	0	0	0	1	1	0	0	0x8C	0	1	1	1	0	0	1	1	0x73
R	1	1	0	0	1	1	1	0	0xCE	0	0	1	1	0	0	0	1	0x31
U	1	1	0	0	0	0	0	1	0xC1	0	0	1	1	1	1	1	0	0x3E
Y	1	0	0	1	0	0	0	1	0x91	0	1	1	0	1	1	1	0	0x6E
—	1	0	1	1	1	1	1	1	0xBF	0	1	0	0	0	0	0	0	0x40
off	1	1	1	1	1	1	1	1	0xFF	0	0	0	0	0	0	0	0	0x00

1.2 Digital Tube Static Display

There are two modes of LED display: static display mode and dynamic display mode. Static display feature is that each digital tube segment selection must be connected with an 8-bit data line to maintain the display of font code. When the font code is sent once, the display font can be maintained until the new font code is sent. This method has the advantage of less CPU time and easy display for monitoring and controlling. The disadvantage is that the hardware circuit is more complex and the cost is higher. (See Figure 2.1.5)

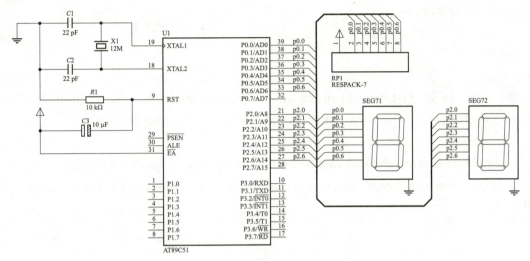

Figure 2.1.5　Two digital tube static display circuit diagram

1.3　The Basic of C Language: One Dimensional Array

Scan the QR code to view the teaching note of One Dimensional Array Concept　　　Scan the QR code to watch the teaching video of One Dimensional Array Concept

The C language provides a capability that enables you to define a set of ordered data items known as an array. This chapter describes how arrays can be defined and manipulated.

Defining an Array

You can define a variable called tab, which represents not a single value of a font encoding, but an entire set of grades. Each element of the set can then be referenced by means of a number called an index number or subscript. Whereas in mathematics a sub-scripted variable x[i] refers to the ith element x in a set, in C the equivalent notation is as follows:

x[i]

So the expression

tab[5]

(read as "tab sub 5") refers to element number 5 in the array called tab. Array elements begin with the number zero, so

tab[0]

actually refers to the first element of the array. (For this reason, it is easier to think of it as referring to element number zero, rather than as referring to the first element.)

An individual array element can be used anywhere that a normal variable can be used. For example, you can assign an array value to another variable with a statement as the following:

g = tab[50];

This statement takes the value contained in tab[50] and assigns it to g. More generally, if i is declared to be an integer variable, the statement

g = tab[i];

takes the value contained in element number i of the grades array and assigns it to g. So if i is equal to 7 when the preceding statement is executed, the value of tab[7] is assigned to g.

A value can be stored in an element of an array simply by specifying the array element on the left side of an equal sign. In the statement

tab[100] = 95;

the value 95 is stored in element number 100 of the tab array. The statement

grades[i] = g;

has the effect of storing the value of g in tab[i].

The capability to represent a collection of related data items by a single array enables you to develop concise and efficient programs. For example, you can easily sequence through the elements in the array by varying the value of a variable that is used as a sub script in the array. So the for loop

```
for ( i = 0; i < 100; ++i )
    sum += tab[i];
```

sequences through the first 100 elements of the array tab (elements 0 to 99) and adds the value of each grade into sum. When the for loop is finished, the variable sum then contains the total of the first 100 values of the tab array (assuming sum was set to zero before the loop was entered).

When working with arrays, remember that the first element of an array is indexed by zero, and the last element is indexed by the number of elements in the array minus one. In addition to integer constants, integer-valued expressions can also be used inside the brackets to reference a particular element of an array.

Just as with variables, arrays must also be declared before they are used. The declaration of an array involves declaring the type of element that will be contained in the array—such as int, float, or char—as well as the maximum number of elements that will be stored inside the array. (The C compiler needs this latter information to determine how much of its memory space to reserve for the particular array.)

As an example, the declaration

```
int tab[100];
```

declares tab to be an array containing 100 integer elements. Valid references to this array can be made by using subscripts from 0 through 99. But be careful to use valid subscripts because C does not do any checking of array bounds for you. So a reference to element number 150 of array grades, as previously declared, does not necessarily cause an error but does most likely cause unwanted, if not unpredictable, program results.

To declare an array called averages that contains 200 floating-point elements, the declaration

    float averages[200];

is used. This declaration causes enough space inside the computer's memory to be reserved to contain 200 floating-point numbers. Similarly, the declaration

    int values[10];

reserves enough space for an array called values that could hold up to 10 integer numbers.

Initializing Arrays

Just as you can assign initial values to variables when they are declared, so can you assign initial values to the elements of an array. This is done by simply listing the initial values of the array, starting from the first element. Values in the list are separated by commas and the entire list is enclosed in a pair of braces.

    The statement
    int counters[5] = { 0, 0, 0, 0, 0 };

declares an array called counters to contain five integer values and initializes each of these elements to zero. In a similar fashion, the statement

    int integers[5] = { 0, 1, 2, 3, 4 };

sets the value of integers[0] to 0, integers[1] to 1, integers[2] to 2, and so on.

Arrays of characters are initialized in a similar manner; thus the statement

    char letters[5] = {'a', 'b', 'c', 'd', 'e'};

defines the character array letters and initializes the five elements to the characters 'a', 'b', 'c', 'd', and 'e', respectively.

It is not necessary to completely initialize an entire array. If fewer initial values are specified, only an equal number of elements are initialized. The remaining values in the array are set to zero. So the declaration

float sample_data[500] = { 100.0, 300.0, 500.5 };

initializes the first three values of sample_data to 100.0, 300.0, and 500.5, and sets the remaining 497 elements to zero.

By enclosing an element number in a pair of brackets, specific array elements can be initialized in any order. For example,

float sample_data[500] = { [2] = 500.5, [1] = 300.0, [0] = 100.0 };

initializes the sample_data array to the same values, as is shown in the previous example.

And the statements

int x = 1233;
int a[10] = { [9] = x + 1, [2] = 3, [1] = 2, [0] = 1 };

define a 10-element array and initialize the last element to the value of x + 1 (or to 1234), and the first three elements to 1, 2, and 3, respectively.

Unfortunately, C does not provide any shortcut mechanisms for initializing array elements. That is, there is no way to specify a repeat count, so if it were desired to initially set all 500 values of sample_data to 1, all 500 would have to be explicitly spelled out. In such a case, it is better to initialize the array inside the program using an appropriate for loop.

1.4 The Basic of C Language: Selection Statement—Switch

The type of if-else statement chain that you encountered in the last program example—in which the value of a variable is successively compared against different values—is so commonly used when developing programs that a special program statement exists in the C language for performing precisely this function. The name of the statement is the switch statement, and its general format is

switch (expression)
{
    case value 1:

```
            program statement
            program statement
            ...
            break;
        case value 2:
            program statement
            program statement
            ...
            break;
        ...
        case value n:
            program statement
            program statement
            ...
            break;
        default:
            program statement
            program statement
            ...
            break;
    }
```

The expression enclosed within parentheses is successively compared against the values value1, value 2, ..., value n, which must be simple constants or constant expressions. If a case is found whose value is equal to the value of expression, the program statements that follow the case are executed. Note that when more than one such program statement is included, they do not have to be enclosed within braces.

The break statement signals the end of a particular case and causes execution of the switch statement to be terminated. Remember to include the break statement at the end of every case. Forgetting to do so for a particular case causes program execution to continue into the next case whenever that case gets executed.

The special optional case called default is executed if the value of expression does not match any of the case values. This is conceptually equivalent to the "fall through" else that you used in the previous example. In fact, the general form of the switch statement can be equivalently expressed as an if statement as follows:

```
if ( expression == value 1)
{
    program statement
    program statement
    ...
}
else if ( expression == value 2)
{
    program statement
    program statement
    ...
}
...
else if ( expression == value n)
{
    program statement
    program statement
    ...
}
else
{
    program statement
    program statement
    ...
}
```

Bearing this mind, you can translate the big if statement into an equivalent switch statement, as is shown as follows.

```
/* Program to evaluate simple expressions of the form value operator value */
#include <stdio.h>
int main (void)
{
    float value1, value 2;
    char operator;
```

```c
        printf ("Type in your expression.\n");
        scanf ("%f %c %f", &value1, &operator, &value2);
        switch (operator)
        {
            case '+':
                printf ("%.2f\n", value1 + value2);
                break;
            case '-':
                printf ("%.2f\n", value1 - value2);
                break;
            case '*':
                printf ("%.2f\n", value1 * value2);
                break;
            case '/':
                if ( value2 == 0 )
                    printf ("Division by zero.\n");
                else
                    printf ("%.2f\n", value1 / value2);
                break;
            default:
                printf ("Unknown operator.\n");
                break;
        }
        return 0;
}
```

After the expression has been read in, the value of operator is successively compared against the values as specified by each case. When a match is found, the statements contained inside the case are executed. The break statement then sends execution out of the switch statement, where execution of the program is complete. If none of the cases match the value of operator, the default case, which displays unknown operator is executed.

The break statement in the default case is actually unnecessary in the preceding program because no statements follow this case inside the switch. Nevertheless, it is a good programming habit to remember to include the break at the end of every case.

When writing a switch statement, bear in mind that no two case values can be the same. However, you can associate more than one case value with a particular set of program statements. This is done simply by listing the multiple case values (with the keyword case before the value and the colon after the value in each case) before the common statements that are to be executed. For example, in the following switch statement, the printf statement, which multiples value1 by value2, is executed if the operator is equal to an asterisk or to the lowercase letter x.

```
switch (operator)
{
    ...
    case '*':
    case 'x':
        printf ("%.2f\n", value1 * value2);
        break;
    ...
}
```

Circuit Diagram

Figure 2.1.6 shows the circuit diagram of independent button access control system.

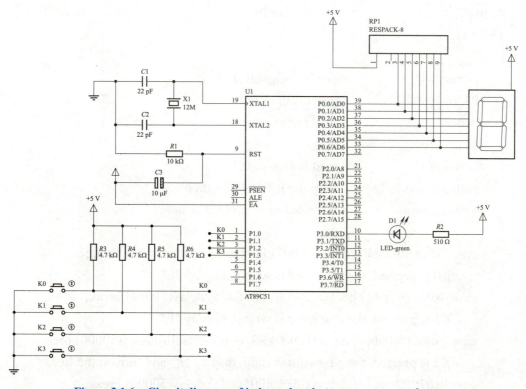

Figure 2.1.6 Circuit diagram of independent button access control system

Program Description

```
/************************************************************************
PROJECT: Independent button access control system design
FILE: PROJ2_1.C
PROCESSOR: AT89C51
This project uses 4 independent buttons to control 7-segment display driver.
************************************************************************/
#include <reg51.h>
sbit P3_0=P3^0;
void delay(unsigned int i);    //declare delay function

/* start of main program */
void main()
{
    unsigned char button;         //save button state
    unsigned char code tab[7]={0xc0,0xf9,0xa4,0xb0,0xbf,0x86,0x8c};
      //corresponding to display characters
    P1=0xff;                      //initialize P1
    while(1)
    {
        P0=tab[4];              //display original state "-"
        P3_0=1;                 //turn off the LED
        button=P1;              //read button state
        delay(1200);            //delay
        button=P1;              //read button state
        button&=0x0f;           //keep the lower 4 bits state
        switch (button)         //if the button is pressed or not
        {
          case 0x0e: P0=tab[0];delay(10000);P0=tab[5];delay(50000);break;
              //K0 is pressed, the password is wrong, display "E"
          case 0x0d: P0=tab[1];delay(10000);P0=tab[5];delay(50000);break;
              //K1 is pressed, the password is wrong, display "E"
          case 0x0b: P0=tab[2];delay(10000);P3_0=0;P0=tab[6];delay(50000);break;
              //K2 is pressed, the password is right, display "P" and turn on the LED
          case 0x07: P0=tab[3];delay(10000);P0=tab[5];delay(50000);break;
              //K3 is pressed, the password is wrong, display "E"
```

```
            }
        }
}

/*Function to delay*/
void   delay(unsigned int i)
{
    unsigned int k;
    for(k=0;k<i;k++);
}
```

Quiz

1. For the same display character, such as "0", there is a relationship between the display code of the common anode digital tube and the common cathode digital tube. (   )
   A) The bitwise reverse.
   B) The bitwise and.
   C) The bitwise or.

2. In common anode digital tube, if only the decimal point is displayed, the corresponding field code is (   ).
   A) 80H
   B) 10H
   C) 40H
   D) 7FH

3. Which program structure do we often use to handle arrays? (   )
   A) Sequential structure.
   B) Selective structure.
   C) Branch structure.
   D) Loop structure.

4. The following statement initializes the one-dimensional array a correctly is (   ).
   A) int a[10]=(0,0,0,0,0);
   B) int a[10]={};
   C) int a[]={0};
   D) int a[10]="10*1";

5. int a [10]={1, 2, 3, 4, 5}; Which of the following options is described correctly? (   )
   A) The initial values of 10 array elements are assigned.
   B) Assign the initial values of the first five array elements.
   C) The initial value of a[9] is 10.
   D) The initial value of a[4] is 5.

Scan the QR code to view the answer of quiz

Summary

Through the independent button access control system design, let the reader understand the application of the array in C language, and understand the interface circuit design and programming control method of MCU and LED digital tubes.

Task 2　Matrix Keyboard Access Control System Design

Objectives

After completing this task, you should be able to
- interface with matrix keyboard;
- use matrix keyboard to input messages.

Task Requirement

Use matrix keyboard to design an access control system. When press a key, the display tube will show the number or character.

2.1　MCU and Matrix Keyboard Interface

Scan the QR code to view the teaching note of Matrix Keyboard

Scan the QR code to watch the teaching video of Matrix Keyboard

The predominant interface between humans and computers is the keyboard. These range in complexity from the "up-down" buttons used for elevators to the personal computer QWERTY layout, with the addition of function keys and numeric keypads. One of the first mass uses for the microcontroller was to interface between the keyboard and the main processor in personal computers. Industrial and commercial applications fall somewhere in between these extremes, using layouts that might feature from six to twenty keys.

The one constant in all keyboard applications is the need to accommodate the human user. Human beings may have little tolerance for machine failure. Watch what happens when the product isn't ejected from the vending machine. Sometimes they are bored or even hostile towards the machine. The hard ware designer has to select keys that will survive in the intended environment and the programmer must write code that will anticipate and defeat inadvertent and also deliberate attempts by humans to confuse the program. It is very important to give instant feedback to the user that the key hit has been acknowledged by the program. By the light a light, beep a beep display the key hit, or whatever, the human user must know that the key has been recognized. Even feedback sometimes is not enough, note the behavior of people at an elevator. Even if the "up" light is lit when we arrive, we will push it again to let the machine know that "I'm here too."

Human Factors

The keyboard application program must guard against the following possibilities:

- More than one key pressed (simultaneously or released in any sequence);
- Key pressed and held;
- Rapid key press and release.

All of these situations can be addressed by hardware or software. Software which is the most cost effective, is emphasized here.

Key Switch Factors

The universal key characteristic is the ability to bounce: The key contacts vibrate open and close for a number of milliseconds when the key is hit and often when it is released.

These rapid pulses are not discernable to humans, but they last a relative eternity in the microsecond-dominated life of the microcontroller. Keys may be purchased that do not bounce, keys may be debounced with RS flip-flops, or debounced in software with time delays.

Keyboard Configurations

The X-Y matrix, keyboard connections shown in Figure 2.2.1 are very popular when the number of keys exceeds ten. The matrix is the most efficient when arranged as a square so that N leads for X and N leads for Y can be used to sense as many as n^2 keys. Matrixes are the most cost effective for large number of keys.

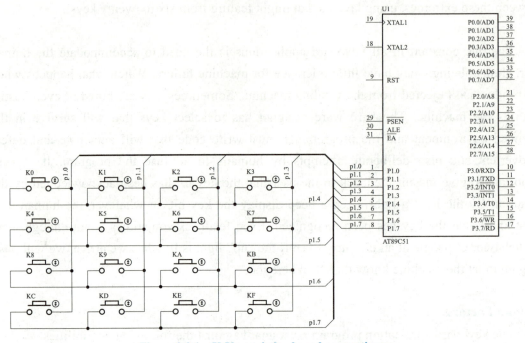

Figure 2.2.1 X-Y matrix keyboard connections

Programs for Keyboards

Programs that deal with humans via keyboards identified in the following manners:

• Bounce: A time delay that is known to exceed the manufacturer's specification is used to wait out the bounce period in both directions.

• Multiple keys: Only patterns that are generated by a valid key pressed are accepted—all others are ignored—and the first valid pattern is accepted.

• Key held: Valid key patterns are accepted after valid debounce delay; no additional keys are accepted until all keys are seen to be up for a certain period of time.

• Rapid key hit: The design is so efficient that the keys are scanned at a rate faster than any human reaction time.

The last item brings up an important point: Should the keyboard be read as the program loops (software polled) or read only when a key has been hit (interrupt driven)?

In general, matrix keyboards are scanned by bringing each X row low in sequence and detecting a Y column low to identify each key in the matrix. X-Y scanning can be done by using dedicated keyboard scanning circuitry or by using the microcontroller ports under program control. The scanning circuitry adds cost to the system. The programming approach takes processor time, and the possibility exists that response to the user may be sluggish if the program is busy elsewhere when a key is hit. Note how long your personal computer takes to respond to a break key when it is executing a print command for instance. The choice between adding scanning hardware or program software is decided by how busy the processor is and the volume of entries by the user.

Circuit Diagram

The circuit diagram is shown in Figure 2.2.2.

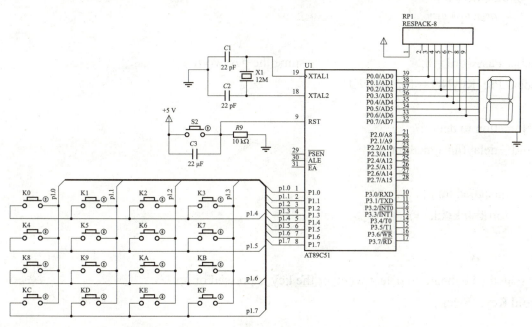

Figure 2.2.2   Circuit diagram

Program Description

/**

PROJECT: Independent button access control system design

FILE: PROJ2_2.C

PROCESSOR: AT89C51

This project uses matrix keyboard to control 7-segment display driver.

**/

#include <reg52.h>

```c
#define uchar unsigned char
#define uint unsigned int
sbit LED = P3^0;

uchar code DSY_CODE[ ]=
{
   0xc0,0xf9,0xa4,0xb0,0x99,0x92,0x82,0xf8,0x80,0x90,0x88,0x83,0xc6,0xa1,0x86,0x8e
};      //0~9,A~F

uchar code DSY_CODE1[ ]=
{
   0xbf,0x8c,0x86
};      //"-","P","E"

uchar Password = 8;                //assume the password is 8
uchar Pre_KeyNO = 16, KeyNO = 16;

/*Function to delay*/
void   delay(unsigned int i)
{
    unsigned int k;
    for(k=0;k<i;k++);
}

/*scan the keyboard to detect whether the key is pressed or not*/
void Keys_Scan()
{
    uchar Tmp;
    P1 = 0x0f;
    delay(100);
    Tmp = P1 ^ 0x0f;
    switch(Tmp)
    {
        case 1: KeyNO = 0; break;
        case 2: KeyNO = 1; break;
        case 4: KeyNO = 2; break;
        case 8: KeyNO = 3; break;
```

```
            default: KeyNO = 16;
    }
    P1 = 0xf0;
    delay(100);
    Tmp = P1 >> 4 ^ 0x0f;
    switch(Tmp)
    {
        case 1: KeyNO += 0; break;
        case 2: KeyNO += 4; break;
        case 4: KeyNO += 8; break;
        case 8: KeyNO += 12;
    }
}

/* start of main program */
void main()
{
    P0 = ~DSY_CODE1[0];;
    while(1)
    {
        P1 = 0xf0;
        if(P1 != 0xf0)           //the key is pressed
            Keys_Scan();   //call the key scan function

        if(Pre_KeyNO != KeyNO)
        {
            if(Password != KeyNO)                //if the password is wrong
            {
                P0 = ~DSY_CODE[KeyNO];       //dispaly the number
                delay(30000);
                P0 = ~DSY_CODE1[2];          //display "E"
                delay(50000);
                P0 = ~DSY_CODE1[0];          //display "-"
            }

            else                                 //if the password is right
            {
                P0 = ~DSY_CODE[KeyNO];       //dispaly the number
```

MCU Practical Technology

```
                delay(30000);
                P0 = ~DSY_CODE1[1];           //display "P"
                delay(50000);
                P0 = ~DSY_CODE1[0];           //display "-"
            }

            Pre_KeyNO = KeyNO;
        }
    }
}
```

Quiz

1. The main working mode of the matrix keyboard is (   ).
    A) programming scan mode and interrupt scan mode
    B) independent query mode and interrupt scan mode
    C) interrupt scanning mode and direct access mode
    D) direct input mode and direct access mode

2. 4*4 matrix keyboard requires 16 data lines to connect. (   )
    A) True.
    B) False.

3. The keyboard can be divided into independent connection and (   ) according to the connection mode.

4. What is the connection method and working principle of the matrix keyboard?

5. What are two common connections to a keyboard?

Scan the QR code to view the answer of quiz

Summary

The matrix keyboard is often used when the keys are more than 10. The scan keyboard code is a little bit complex. The way to study this part is to read the code more than three times, and try to remember. When you read and remember several codes like this, you will realize you can write it by yourself.

Task 3   Independent Button Access Control System Design with Interrupt

Objectives

After completing this task, you should be able to
- understand the concept of MCU interrupt;
- explain special function register of MCU interrupt;
- write interrupt service routines.

Task Requirement

The first two tasks use independent button or matrix keyboard to design access control system. However, in general case, the MCU will not stay waiting until the button is pressed, it often processes other things while waiting. It introduces interrupt. An interrupt is something that stops what is happening to let something else happen. You can imagine eating dinner and being interrupted by a telephone call. The same thing can happen with the 8051 family of processors. A program can be running when some signal comes in that makes the processor switch over to run another program. For a computer, an interrupt is a hardware- related event that stops the processor at the place in the program where it is running and causes it to move to some other place in the program.

When the key is pressed, the 7-segment digital tube displays 'P', and the LED is turned on. When the key is released, the 7-segment digital tube displays '−', and the LED is turned off.

The key is connected to P3.2, as an interrupt 0 input.

3.1   MCU Interrupt System

Scan the QR code to view the teaching note of MCU Interrupt System

Scan the QR code to watch the teaching video of MCU Interrupt System

A computer program has only two ways to determine the conditions that exist in internal and external circuits. One method uses software instructions that jump on the states of flags and port pins. The second responds to hardware signals, called interrupts that force the program to call a sub-routine. Software techniques use up processor time that can be devoted to other tasks; interrupts take processor time only when an action is needed by the program. Most applications of microcontrollers involve responding to events quickly enough to control the environment that generates the events (generically termed "real-time programming"). Interrupts are often the only way in which real-time programming can be done successfully.

Interrupts may be generated by internal chip operations or provided by external sources. Any interrupt can cause the 8051 to perform a hardware call to an interrupt-handling subroutine that is located at a predetermined (by the 8051 designers) absolute address in program memory.

Five interrupts are provided in the 8051. Three of these are generated automatically by internal operations: timer flag 0, timer flag 1, and the serial port interrupt (RI or TI). Two interrupts are triggered by external signals provided by circuitry that is connected to pins INT0 and INT1 (port pins P3.2 and P3.3).

All interrupt functions are under the control of the program. The programmer is able to alter control bits in the interrupt enable control register (IE), the interrupt priority control register (IP) and the timer/counter control register (TCON). The program can block all or any combination of the interrupts from acting on the program by suitably setting or clearing bits in these registers. The IE and IP registers are shown in Table 2.3.1 and Table 2.3.2.

After the interrupt has been handled by the interrupt subroutine which is placed by the programmer at the interrupt location in program memory, the interrupted program must resume operation at the instruction where the interrupt took place. Program resumption is done by storing the interrupted PC address on the stack in RAM before changing the PC to the interrupt address in ROM. The PC address will be restored from the stack after an RETI instruction is executed at the end of the interrupt subroutine.

The standard 8051 provides six interrupt sources as follows:

- Two external interrupts ($\overline{INT0}$ and $\overline{INT1}$);
- Two timer interrupts (timer 0 and timer 1);
- One serial port receive interrupt (RI);

- One serial port transmit interrupt (PI).

Each interrupt is assigned a fixed location in memory and an interrupt causes the CPU to jump to that location where it executes the interrupt service routine. Table 2.3.3 gives the interrupt sources and the start of their service routines in memory. Note that the serial port receive and transmit interrupts point to the same location.

Each interrupt source can be individually enabled or disabled by setting or clearing its interrupt enable bit. Table 2.3.1 gives the interrupt enable bit patterns.

Table 2.3.1 Interrupt enable control register (IE)

EA	—	—	ES	ET1	EX1	ET0	EX0

Where:
EA: Global interrupt enable/disable. If EA = 0, no interrupt will be accepted. If EA = 1, each interrupt source is individually enabled or disabled by setting or clearing its bit given below.
ES: Serial port interrupt enable bit.
ET1: Timer 1 interrupt enable bit.
EX1: External interrupt 1 enable bit.
ET0: Timer 0 interrupt enable bit.
EX0: External interrupt 0 enable bit

Table 2.3.2 Interrupt priority control register (IP)

—	—	—	PS	PT1	PX1	PT0	PX0

Where:
PS: Priority of serial port interrupt. Set/cleared by program.
PT1: Priority of timer 1 overflow interrupt. Set/cleared by program.
PX1: Priority of external interrupt 1. Set/cleared by program.
PT0: Priority of timer 0 overflow interrupt. Set/cleared by program.
PX0: Priority of external interrupt 0. Set/cleared by program

Table 2.3.3 Interrupt entry locations in memory

Interrupt source	Interrupt number	Location in memory (hex)
External interrupt 0	0	0003
Timer 0	1	000B
External interrupt 0	2	0013
Timer 1	3	001B
Serial port	4	0023

External Interrupts

Pins $\overline{INT0}$ and $\overline{INT1}$ are used by external circuitry. Inputs on these pins can set the interrupt flags IE0 and IE1 in the TCON register to 1 by two different methods. The IEX flags may be set when the $\overline{INTX}$ pin signal reaches a low level, or the flags may be set when a high-to-low transition takes place on the $\overline{INTX}$ pin. Bits IT0 and IT1 in TCON program the $\overline{INTX}$ pins for low-level interrupt when set to 0 and program the $\overline{INTX}$ pins for transition interrupt when set to 1.

Flags IEX will be reset when a transition-generated interrupt is accepted by the processor and the interrupt subroutine is accessed. It is the responsibility of the system signer and programmer to reset any level-generated external interrupts when they are serviced by the program. The external circuit must remove the low level before an RETI is executed. Failure to remove the low will result in an immediate interrupt after RETI, from the same source.

Interrupt Control

The program must be able, at critical times, to inhibit the action of some or all of the interrupts so that crucial operations can be finished. The IE register holds the programmed bits that can enable or disable all the interrupts as a group, or if the group is enabled, each individual interrupt source can be enabled or disabled.

Often, it is desirable to be able to set priorities among competing interrupts that may conceivably occur simultaneously. The IP register bits may be set by the program to assign priorities among the various interrupt sources so that more important interrupts can be serviced first? should two or more interrupts occur at the same time.

Interrupt Enable/Disable

Bits in the IE register are set to 1 if the corresponding interrupt source is to be enabled and set to 0 to disable the interrupt source. Bit EA is a master, or "global" bit that can enable or disable all of the interrupts.

Interrupt Priority

Register IP bits determine if any interrupt is to have a high or low priority. Bits set to 1 give the accompanying interrupt a high priority while a 0 assigns a low priority. Interrupts with a high priority can interrupt another interrupt with a lower priority; the low priority interrupt continues after the higher is finished.

If two interrupts with the same priority occur at the same time, then they have the

following ranking:
1. IE0;
2. TF0;
3. IE1;
4. TF1;
5. Serial = RI or TI.

The serial interrupt can be given the highest priority by setting the PS bit in IP to 1, and all others to 0.

Interrupt Destinations

Each interrupt source causes the program to do a hardware call to one of the dedicated addresses in program memory. It is the responsibility of the programmer to place a routine at the address that will service the interrupt.

The interrupt saves the PC of the program, which is running at the time the interrupt is serviced on the stack in internal RAM. A call is then done to the appropriate memory location. These locations are shown in Table 2.3.3.

A RETI instruction at the end of the routine restores the PC to its place in the interrupted program and resets the interrupt logic so that another interrupt can be serviced. Interrupts that occur but are ignored due to any blocking condition (IE bit not set or a higher priority interrupt already in process) must persist until they are serviced, or they will be lost. This requirement applies primarily to the level-activated INTX interrupts.

Software Generated Interrupts

When any interrupt flag is set to 1 by any means, an interrupt is generated unless blocked. This means that the program itself can cause interrupts of any kind to be generated simply by setting the desired interrupt flag to 1, using a program instruction.

3.2 MCU Interrupt Processing Function

The C51 compiler allows us to declare interrupt service routines (ISRs) in our C code and then the program automatically jumps to this code when an interrupt occurs. The compiler automatically generates the interrupt vectors and entry and exit code for interrupt routines.

An ISR is declared similar to a function declaration but the interrupt number is specified as part of the function declaration. For example, the following is a declaration of the ISR for timer 1 interrupts (interrupt number 3):

```
void timer1() interrupt 3
{
    //interrupt service code goes in here
}
```

Similarly, the ISR for timer 0 (interrupt number 1) is declared as:

```
void timer0() interrupt 1
{
    //interrupt service code goes in here
}
```

Note that we can specify the register bank to be used for the ISR with the using function attribute:

```
void timer0() interrupt 1 using 2
{
    //interrupt service code goes in here
}
```

Circuit Diagram

The circuit diagram is shown in Figure 2.3.1.

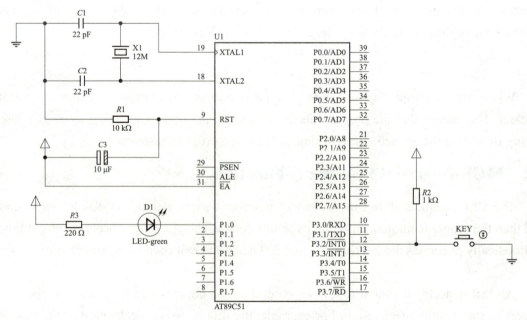

Figure 2.3.1　Circuit diagram

Program Description

```
/*********************************************************************
PROJECT: Independent button access control system design with interrupt
FILE: PROJ2_3.C
PROCESSOR: AT89C51
This project uses an independent button to control LED.
*********************************************************************/
#include <reg51.h>

sbit P1_0=P1^0;
unsigned char led;

/* start of main program */
void main()
{
    EA =1;       //enable interrupt
    EX0 = 1;     //enable external interrupt
    IT0 = 1;
    led = 0;
    while(1)
    {
       P1_0=led;
    }
}

/*Interrupt function*/
void int_0() interrupt 0
{
    led = ~led;
}
```

Quiz

1. What is the place where the main program is broken? (   )
   A) The interrupt source.
   B) Entrance address.
   C) Interrupt vector.

D) A breakpoint.

2. MCU connected to all external equipments can be an interrupt source. (   )
   A) True.
   B) False.

3. How many interrupt sources does MCU interrupt system have? (   )
   A) 1.
   B) 2.
   C) 4.
   D) 5.

4. How many interrupt priority levels does MCU interrupt system have? (   )
   A) 1.
   B) 2.
   C) 4.
   D) 5.

5. Which of the following is the global interrupt allowed control bit? (   )
   A) ES.
   B) ET1.
   C) EX1.
   D) EA.

Scan the QR code to view the answer of quiz

Summary

This project designs household access control system. The main points of this project are digital tube display, maxtrix keyboard and external interruption. Students are required to master the writing of interrupt handling functions.

Extended Reading

The development of Chips in China

In February 1955, during the first Five-Year Plan Period, Peking University offered China's first semiconductor course, led by Huang Kun, Yang Zhenning's university friend and one of the world's leading physicists, and Xie Xide, one of Fudan's top physicists who later became the university's president.

In 1957, Beijing Electron Tube Factory pulled out China's first single crystal of germanium. In the same year, Wang Shouwu and Wang Shoujue, the brother scientists developed China's earliest semiconductor device—germanium alloy transistor.

Considering that the foundation was extremely weak at that time, so this achievement was quite good.

In September 1965, with the joint efforts of Shanghai Metallurgical Research Institute and Shanghai Component No.5 Factory, China's first integrated circuit was successfully developed ahead of the Semiconductor Research Institute in Beijing.

That is seven years behind the United States, on a par with Japan, and 10 years ahead of the Republic of Korea.

In 1966, Nantong Transistor Factory, the predecessor of Tongfu Microelectronics was established.

In 1968, the state-owned Dongguang Power Plant and Radio 19 Plant were set up in Beijing and Shanghai respectively. Sijibu also set up 1424 Institute in Yongchuan, Chongqing.

In 1969, Sijibu established the predecessor of Huatian Technology, Gansu Tianshui Yonghong Equipment Factory and Gansu Tianshui Tianguang integrated Circuit Factory.

On December 13, 1995, the State Council launched Project 909 which then became the largest national project in the history of China's electronics industry. 909 gave birth to Huahong. Since the reform, there have been numerous sino-foreign joint ventures in China, and Huahong NEC is undoubtedly the most successful representative.

In 2003, Huahong NEC was officially upgraded from a semiconductor processing plant to the first fab in China when the Japanese side withdrew from the company.

In October 2004, Huawei Hisilicon was established, formerly Huawei IC Design Center established in 1991. In the field of mobile phone chips, Hisilicon is currently the only one in China in the field of smart phones.

In June 2014, the Outline of promoting the "Development of The National IC Industry" was officially released, which elevated the development of the IC industry to a national strategy, defined the focus and goals of the development of the domestic IC industry during the 13th Five-Year Plan period, and opened a new stage of the development of China's IC industry. In 2014, the scale of China's IC design industry exceeded 100 billion yuan. The scale of China's integrated circuit industry broke through 300 billion yuan for the first time, and it took two years from 200 billion to 300 billion yuan.

In 2014, Hisilicon launched 4G mobile phone chip Kirin 920, which has stable performance and has become an important force in mobile phone chips.

Until 2020, Kirin chip products have been updated to the Kirin 990 in 5G high-end models, and Hisilicon has also ushered in its peak, forming Kirin series, Kunpeng series and Shengteng series. Baseband products Tiangang and Balong and other special chips.

Project 3

Digital Clock Control System Design

Task 1　Simple Stopwatch Design

Objectives

After completing this task, you should be able to
- understand the concept of MCU timer/counter;
- explain the working mode and working process of MCU timer/counter;
- create time delays using the timer function.

Task Requirement

Through the design and manufacture of the simple stopwatch control system displayed by 2 LED digital tubes, students should be familiar with the timer/counter and interrupt programming control method of MCU, including the setting of timer/counter working mode, setting of initial value, interrupt programming and application of interrupt function.

Control 2 LED digital tubes with MCU, using static connection, 2 digital tubes display 00–99 count, the time interval is 1s.

1.1　Timer/Counters' Structure and Working Mode

Scan the QR code to view the teaching note of Timer/Counter Introduction

Scan the QR code to watch the teaching video of Timer/Counter Introduction

Many microcontroller applications require the counting of external events, such as the frequency of a pulse train, or the generation of precise internal time delays between computer actions. Both of these tasks can be accomplished using software techniques, but software loops for counting or timing keep the processor occupied so that other, perhaps more important functions are not done. To relieve the processor of this burden two 16-bit up counters, named T0 and T1 are provided for the general use of the programmer. Each counter may be programmed to count internal clock pulses, acting as a timer, or programmed to count external pulses as a counter.

The 8051 contains two timers/counters known as timer/counter 0 and timer/counter 1 (larger members of the 8051 family contain more timers/counters). The timers/counters are divided into two 8-bit registers called the timer low (TL0, TL1) and high (TH0, TH1) bytes. These timers/counters can be operated in several different modes depending upon the programming of two registers TCON and TMOD, as is shown in Tables 3.1.1 and Tables 3.1.2. TMOD is dedicated solely to the two timers and can be considered to be two duplicate 4-bit registers, each of which controls the action of one of the timers. TCON has control bits and flags for the timers in the upper nibble, and control bits and flags for the external interrupts in the lower nibble. TCON is the timer/counter control register and this register is bit addressable. These registers should be programmed before using any timer or counter facilities of the microcontroller.

Table 3.1.1 TCON timer/counter control register

Bit name	Bit position	Description
TF1	7	Timer 1 overflow flag. Set and cleared by hardware
TR1	6	Timer 1 run control bit. Timer 1 is turned on when TR1 = 1, and stopped when TR1 = 0
TF0	5	Timer 0 overflow flag. Set and cleared by hardware
TR0	4	Timer 0 run control bit. Timer 0 is turned on when TR1 = 1, and stopped when TR1 = 0
IE1	3	External interrupt 1 edge flag. Set and cleared by hardware
IT1	2	External interrupt 1 type. IT1 = 1 specifies interrupt on falling edge. IT1 = 0 specifies interrupt on low level
IE0	1	External interrupt 0 edge flag. Set and cleared by hardware
IT0	0	External interrupt 0 type. IT0 = 1 specifies interrupt on falling edge. IT0 = 0 specifies interrupt on low level

Table 3.1.2 TMOD timer/counter mode control register

TIMER 1				TIMER 0			
GATE	C/$\overline{T}$	M1	M0	GATE	C/$\overline{T}$	M1	M0

GATE:	When TRX is set and GATE = 1, timer/counterx runs only while the INTX pin is high. When GATE = 0, timer/counterx will run only while TRX = 1.
C/$\overline{T}$:	Timer or counter select bit. When C/$\overline{T}$ = 0, operates as a timer (from internal clock). When C/$\overline{T}$ = 1, it operates as a counter (input from TX input).
M1, M0:	Timer/counter mode select bits are defined in Table 3.1.3

Table 3.1.3 M1, M0 mode control bits

M1	M0	Operating mode
0	0	13-bit timer
0	1	16-bit timer/counter
1	0	8-bit auto-reload timer/counter
1	1	Two 8-bit timers

For example, bit 4 of TCON is the counter 0 run control bit and setting this bit starts from counter 0. TCON register is at address 88 (hex) and bits in this register can be accessed either by making reference to the address or by using compiler reserved names (e.g. TR0).

TMOD is the timer/counter mode control register. This register sets the operating modes of the two timers/counters. There are 4 operating modes, known as modes 0, 1, 2 and 3. TMOD is not bit addressable and should be loaded by specifying all the 8 bits. For example, loading hexadecimal byte 01 into TMOD sets timer 0 into mode 1 which is a 16-bit timer and is turned on and off by bit TR0 of TCON. Also, timer 1 is set into mode 0 which is a 13-bit timer and is turned on and off by bit TR1 of TCON.

Timing

If a counter is programmed to be a timer, it will count the internal clock frequency of the 8051 oscillator divided by 12d. For example, if the crystal frequency is 6.0 MHz, then the timer clock will have a frequency of 500 KHz.

The resultant timer clock is gated to the timer by means of the circuit shown in Figure 3.1.1. In order that oscillator clock pulses to reach the timer, the C/T bit in the TMOD register must be set too (timer operation). Bit TRX in the TCON register must be set to 1 (timer run), and the gate bit in the TMOD register must be 0, or external pin INTX must be 1. In other words, the counter is configured as a timer then the timer pulses are gated the counter by the run bit and the gate bit or the external input bits INTX.

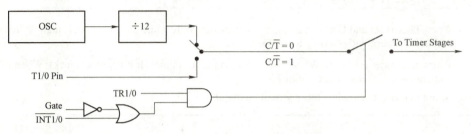

Figure 3.1.1 Timer/counter operating principle

Timer Modes of Operation

The timers may operate in any one of four modes that are determined by the mode bits, M1 and M0, in the TMOD register.

Timer Mode 0

Setting timer X mode bits to 00b in the TMOD register results in using the THX register as an 8-bit counter and TLX as a 5-bit counter. The pulse input is divided by 32d in TL so that TH counts the original oscillator frequency reduced by a total 384d. For example, the 6 MHz oscillator frequency will result in a final frequency to TH of 15,625 Hz. The timer flag is set whenever THX goes from FFh to 00h, or in 0.0164 s for a 6 MHz crystal if THX starts at 00h. (See Figure 3.1.2)

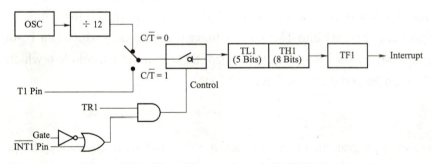

Figure 3.1.2 Timer/Counter 1 mode 0: 13-bit counter

Timer Mode 1

Mode 1 (Figure 3.1.3) is similar to mode 0 except TLX is configured as a full 8-bit counter when the mode bits are set to 01b in TMOD. The timer flag would be set in 0.1311 s using a 6 MHz crystal.

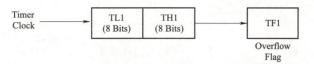

Figure 3.1.3　Timer/Counter 1 mode 1: 16-bit counter

Timer Mode 2

Setting the mode bits to 10b in TMOD configures the timer to use only the TLX counter as an 8-bit counter. THX is used to hold a value that is loaded into TLX every time TLX overflows from FFh to 00h. The timer flag is also set when TLX overflows. (Figure 3.1.4)

This mode exhibits an auto-reload feature: TLX will count up from the number in THX overflow, and be initialized again with the contents of THX. For example, placing 9Ch in THX will result in a delay of exactly 0.000,2 s before the overflow flag is set if a 6 MHz crystal is used.

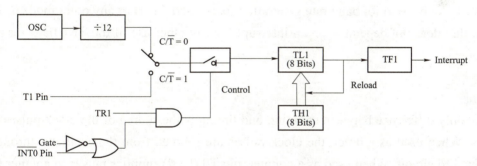

Figure 3.1.4　Timer/Counter 1 mode 2: 8-bit auto-reload

Timer Mode 3

Timers 0 and 1 may be programmed to be in mode 0, 1, or 2 independently of a similar mode for the other timer. This is not true for mode 3—the timers do not operate independently if mode 3 is chosen for timer 0. Placing timer 1 in mode 3 causes it to stop counting. The control bit TR1 and the timer 1 flag TF1 are then used by timer 0. Timer 0 in mode 3 becomes two completely separate 8-bit counters. TL0 is controlled by the gate arrangement of Figure 3.1.5 and sets timer flag TF0 whenever it overflows from FFh to 00h. TH0 receives the timer clock (the oscillator is divided by 12) under the control of TR1 only and sets the TF1 flag when it overflows.

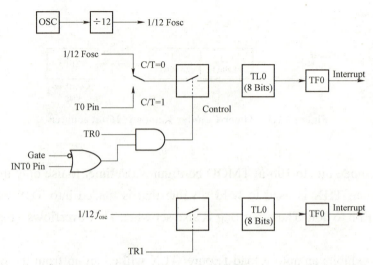

Figure 3.1.5   Timer/Counter 1 mode 3: two 8-bit timers

Timer 1 may still be used in modes 0, 1, and 2, while timer 0 is in mode 3 with one important exception: No interrupts will be generated by timer 1 while timer 0 is using the TF1 overflow flag. Switching timer 1 to mode 3 will stop it (and hold whatever count is in timer 1). Timer 1 can be used for baud rate generation for the serial port or any other mode 0, 1, or 2 function that does not depend upon an interrupt (or any other use of the TF1 flag) for proper operation.

Counting

The only difference between counting and timing is the source of the clock pulses to the counters. When used as a timer, the clock pulses are sourced from the oscillator through the divide-by-12d circuit. When used as a counter, pin T0 (P3.4) supplies pulses to counter 0, and pin T1 (P3.5) to counter 1. The C/T bit in TMOD must be set to 1 to enable pulses from the TX pin to reach the control circuit.

The input pulse on TX is sampled during P2 of state 5 every machine cycle. A change on the input from high to low between samples will increment the counter. Each high and low state of the input pulses must be held constant for at least one machine cycle to ensure reliable counting. Since this takes 24 pulses, the maximum input frequency that can be accurately counted is the oscillator frequency divided by 24. For our 6 MHz crystal, the calculation yields a maximum external frequency of 250 kHz.

Circuit Diagram

The circuit diagram is shown in Figure 3.1.6.

Project 3 Digital Clock Control System Design

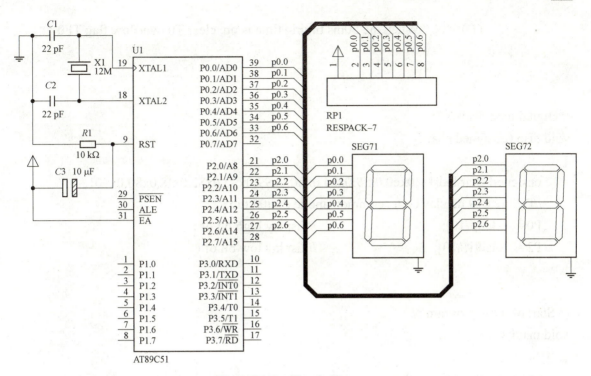

Figure 3.1.6 Circuit diagram

Program Description

```
/********************************************************************
PROJECT: Digital clock control system design
FILE: PROJ3_1.C
PROCESSOR: AT89C51
This project designs a simple stopwatch.
********************************************************************/
#include   "reg51.h"
/*1s delay function*/
void delay1s ( )
{
    unsigned char i;
        for(i=0; i<20; i++)
        {
            TH0 =   (65536-50000)/256;       //Reset the high 8 bits of T0
            TL0 =   (65536-50000)%256;       //Reset the low 8 bits of T0
            TR0= 1;                          //turn on T0
            while(!TF0);                     //Query whether the count is overflowing
```

                TF0 = 0; // When the 50ms timing time is up, clear T0 overflow flag TF0
        }
}

/*Digital tube display*/
void disp (unsigned char i)
{
    unsigned char led[]={0xc0,0xf9,0xa4,0xb0,0x99,0x92,0x82,0xf8,0x80,0x90};
    //Define 0~9 display code, common anode digital tube
    P0=   ~led[i/10];   //display high bit of i
    P2=   ~led[i%10];   //display low bit of i
}

/* Start of main program */
void main()
{
    unsigned char miao = 0;   //second counter definition
    TMOD=   0x01;   //set T0 as mode 1
    TR0 = 1;   //turn on T0
    while(1)
    {
        disp(miao);   //display second counter
        delay1s ( );   //delay 1s
        miao++;   //second counter plus 1
        if(miao == 100) miao = 0;   //if second counter is equal to 100, set it to 0
    }
}

Quiz

1. There are two (   ) programmable timers/counters in 8051 MCU.
   A) 32-bit   B) 16-bit   C) 8-bit   D) 4-bit

2. Control MCU internal timer startup timer is TMOD. (   )
   A) True.   B) False.

3. The counting function of the timer is to count the pulse of the external pin, and the timing function is to count the internal machine cycle. (   )

A) True. B) False.

4. Set T0 as working mode 1, timing function, GATE=0; T1 is working mode 2, counting function, GATE=0. Working mode control register TMOD should be assigned. (   )
   A) 0x20.   B) 0x60.   C) 0x21.   D) 0x61.

5. The following four special registers, the bit addressable is (   ).
   A) SCON   B) TMOD   C) TL0   D) TH0

Scan the QR code to view the answer of quiz

Summary

Through the design and manufacture of the simple stopwatch control system displayed by two LED digital tubes, students should be familiar with the timer/counter and interrupt programming control method of MCU, including the setting of timer/counter working mode, setting of initial value, interrupt programming and application of interrupt function.

Task 2 Design of 24 - hour Digital Clock with Timing Function

Objectives

After completing this task, you should be able to
- understand the interrupt of MCU timer/counter;
- use timer/counter to design a 24-hour digital clock with timing function.

Task Requirement

Design a 24-hour digital clock with the following functions:
- Implement 24-hour digital clock with timer 1;
- Display hour, minute and second information;
- Modify time information;
- Start/stop the clock;

- Reset the clock.

2.1 Timer/Counter Interrupts

Scan the QR code to view the teaching note of Timer/Counter Interrupt

Scan the QR code to watch the teaching video of Timer/Counter Interrupt

The counters have been included on the chip to relieve the processor of timing and counting chores. When the program wishes to count a certain number of internal pulses or external events, a number is placed in one of the counters. The number represents the maximum count, less than the desired count, plus one. The counter increments from the initial number to the maximum and then rolls over to zero on the final pulse and also sets a timer flag. The flag condition may be tested by an instruction to tell the program that the count has been accomplished or the flag may be used to interrupt the program.

Timer Flag Interrupt

When a timer/counter overflows, the corresponding timer flag TF0 or TF1 is set to 1. The flag is cleared to 0 when the resulting interrupt generates a program call to the appropriate timer subroutine in memory.

2.2 Digital Tube Dynamic Display

Scan the QR code to view the teaching note of 7-segment Digital Tube Dynamic Display

Scan the QR code to watch the teaching video of 7-segment Digital Tube Dynamic Display

The use of static display mode to control six digital tubes requires microcontroller to provide six groups of 8-bit parallel I/O ports, which must be extended to microcontroller parallel I/O ports, which will greatly increase the complexity and cost of hardware circuits. This task uses dynamic display mode to control six common anode digital tubes, and the circuit connection mode is shown in Figure 3.2.1. The corresponding segment selection control end of each common anode digital tube is connected in parallel, which is controlled by only one P0 port and driven by eight-phase three-state buffer/line driver 74LS245. The common end of each digital tube is also called 'bit selection end', which is controlled by P2 port and driven by 74LS04, a six-phase inverter driver.

The feature of dynamic display is that all digital tube segment selection is in parallel, by the bit selection control which digital tube is effective. Choose digital tube using dynamic scanning display. The so-called dynamic scanning display is to take turns to send the font code and the corresponding position selection to each digital tube, and the use of luminescent tube afterglow and the role of temporary human vision, make people feel as if each digital tube at the same time is in the display. The brightness of dynamic display is worse than that of static display, so the current limiting resistance should be slightly smaller than that of static display circuit.

Dynamic display means that each digital tube is lit up in turn in order of bits, that is, at a certain period of time, only one of the digital tube "bit selection end" is valid, and send out the corresponding font display code. For example, first let the first digital tube on the left display character 9, P2 port to send out the bit selection code, that is, the statement "P2=0xfe;". After inverting the drive, control P2.0 connected digital tube position selection end is high level, light the digital tube, at the same time, the P0 port sends out "9" font code, the statement "P0=0x90;", digital tube display character 9. Then, use the same method programming in order to display the second to sixth digital tube.

Circuit Diagram

Scan the QR code to view the teaching note of
24-hour Digital Clock Project Analysis

Scan the QR code to watch the teaching video of
24-hour Digital Clock Project Analysis

The circuit diagram is shown in Figure 3.2.1.

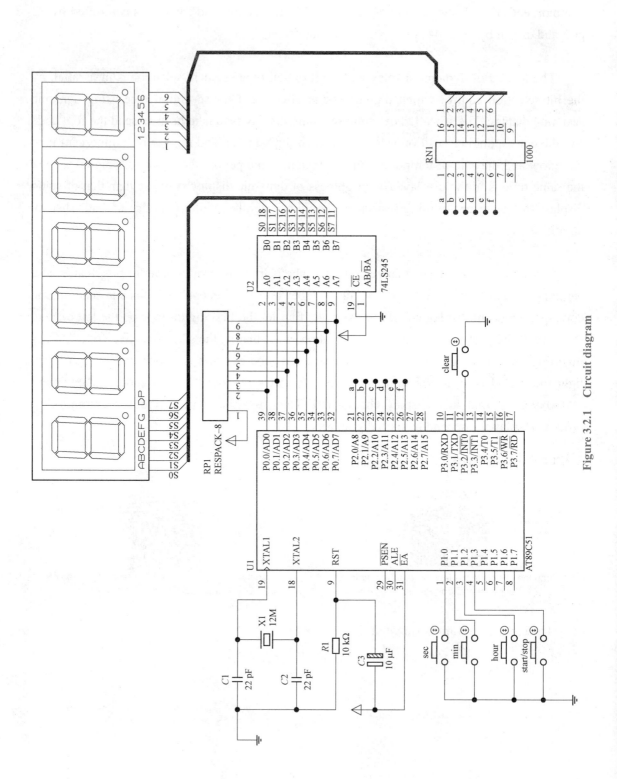

Figure 3.2.1 Circuit diagram

Program Description

```
/********************************************************************
PROJECT: Digital clock control system design
FILE: PROJ3_2.C
PROCESSOR: AT89C51
This project designs a 24-hour digital clock with timing function.
********************************************************************/
#include <reg51.h>
sbit P1_0=P1^0;        //second adjustment button
sbit P1_1=P1^1;        //minute adjustment button
sbit P1_2=P1^2;        //hour adjustment button
sbit P1_3=P1^3;        //run/stop button

unsigned char code DSY_CODE[]=
{
    0x3f,0x06,0x5b,0x4f,0x66,0x6d,0x7d,0x07,0x7f,0x6f
};
unsigned char miao=55, min=59, hour=23, key=0;        //set the initial time value
unsigned char count=0;

/*Function to delay*/
void delay(unsigned char k)
{
    unsigned char i;
    for(i=0;i<k;i++);
}

/*Keyboard scan*/
void keyscan()
{
    key=0;
    if((P1&0x0f)!=0x0f)
    {
        delay(10);
        if((P1&0x0f)!=0x0f)
        {
```

```c
            if(P1_0==0)
                key=1;
            else if(P1_1==0)
                key=2;
            else if(P1_2==0)
                key=3;
            else key=4;
            while((P1&0x0f)!=0x0f);
        }
    }
}

/*Digital tube display*/
void disp()
{
    P2=0xff;        //display off
    P0=DSY_CODE[hour/10];
    P2=0xfe;
    delay(100);

    P2=0xff;        //display off
    P0=DSY_CODE[hour%10]|0x80;
    P2=0xfd;
    delay(100);

    P2=0xff;        //display off
    P0=DSY_CODE[min/10];
    P2=0xfb;
    delay(100);

    P2=0xff;        //display off
    P0=DSY_CODE[min%10]|0x80;
    P2=0xf7;
    delay(100);

    P2=0xff;        //display off
    P0=DSY_CODE[miao/10];
```

```c
        P2=0xef;
        delay(100);

        P2=0xff;       //display off
        P0=DSY_CODE[miao%10];
        P2=0xdf;
        delay(100);
}

/* start of main program */
void main()
{
    TMOD=0x10;
    TH1=(65536-50000)/256;
    TL1=(65536-50000)%256;
    ET1=1;
    EX0=1;
    EA=1;
    IT0=0;
    TR1=0;
    while(1)
    {
        disp();
        keyscan();
        switch(key)
        {
            case 4:TR1=!TR1;break;
            case 1:if(++miao>=60) miao=0;break;
            case 2:if(++min>=60) min=0;break;
            case 3:if(++hour>=24) hour=0;break;
            default:break;
        }
    }
}

void timer_1() interrupt 3
{
```

```
        TH1=(65536-50000)/256;
        TL1=(65536-50000)%256;
        count++;
        if(count==20)          //1s
        {
            count=0;
            miao++;
            if(miao==60)
            {
                miao=0;
                min++;
                if(min==60)
                {
                    min=0;
                    hour++;
                    if(hour==24) hour=0;
                }
            }
        }
}
void int_0() interrupt 0
{
    miao=0;min=0;hour=0;
    TH1=(65536-50000)/256;
    TL1=(65536-50000)%256;
}
```

Quiz

1. The count overflow flag bit of T0 is (   ).
   A) TF0 in TCON   B) TF1 in TCON
   C) TR0 in TCON   D) TR1 in TCON

2. Statements TR1 = 1; The function of (   ).
   A) start T1   B) start T0
   C) stop T1   D) stop T0

3. When the timer T0 counts, (   ) is the mark after overflow?
   A) TR0=0   B) TF0=1
   C) TR0=1   D) TF0=0

4. TR0=1 means start timer T0. (   )
   A) True.   B) False.

5. Timer T0 works in mode 2, if you need to count 50 times, then the initial value is (   ) of the following.
   A) 50   B) 0
   C) 256   D) 206

Scan the QR code to view the answer of quiz

Summary

Through the design and manufacture of digital clock, what we have learned before the MCU internal timer resources, I/O parallel ports, keyboard and display interface achieve mastery through a comprehensive study of knowledge, enhance the ability of independent design, production and debugging application system, grasp MCU application system hardware design, modular program design and the hardware and software debugging methods, and master the development process of MCU application system.

Extended Reading

Global semiconductor industry transfer and industrial chain change

Due to the characteristics of the semiconductor industry, such as extensive downstream applications, multiple production technology processes, multiple product types, rapid technological upgrading, and high investment risk, the continuous rise of the superposition downstream application market has made the semiconductor industry chain change from integration to vertical division of labor more and more clearly, and it has experienced two spatial industrial transfers.

1. Origin, United States, Vertically integrated module

In the 1950s, the semiconductor industry originated in the US and was dominated by system manufacturers. The initial form of the global semiconductor industry was a vertically integrated operation mode, in which all the manufacturing departments of the semiconductor industry were set up within the enterprise to meet the needs of the enterprise's own products.

2. Household appliances, from the USA to Japan, IDM mode

In the 1970s, the United States assembly industry transferred to Japan, and the semiconductor industry came into IDM (Integrated Device Manufacture) model, which is responsible for all the process from design, manufacturing to packaging testing. Unlike vertical integration, IDM's chip products are designed to meet the needs of other system vendors. As the home appliance industry and the semiconductor industry promote each other's development, Japan has incubated SONY, Toshiba and other manufacturers. Most discrete device manufacturers in China also adopt this model.

3. PC (Personal Computer), from the USA and Japan to the Republic of Korea and Chinese Taiwan, OEM mode

In the 1990s, with the rise of the PC, the storage industry shifted from the USA to Japan and then the Republic of Korea, spawning Samsung and Hynix. At the same time, Chinese Taiwan's IC company after its establishment, opened a wafer foundry model, solved the huge investment which is a must to design chip wafer manufacturing production lines, opened the prelude of vertical contract, and the line (Fabless) have set up the design of the company. The traditional IDM vendors such as Intel, samsung in wafer foundry, vertical division of labor pattern gradually became mainstream, formed design (Fabless) to manufacture (Foundry), testing (OSAT) three links.

4. Smart phone, from global to China

In the 2010s, China's smartphone brand global market share continued to improve, which has given rise to the strong demand for semiconductors, combined with the state of the semiconductor industry support. The talent, technology, capital, industrial environment matured, the global semiconductor industry brewed industry transfer, for the third time that transferred trend to China to show itself.

Labor cost is an important driving force for the change and transfer of industrial chain. The integrated circuit industry in the Republic of Korea and Chinese Taiwan started from OEM, and the main factor for OEM choice was labor cost. At that time, the labor cost in the Republic of Korea and Chinese Taiwan was much lower than that in Japan, so the packaging and testing industry began to transfer from Japan to the Republic of Korea and Chinese Taiwan.

Also due to the advantage of human cost, in the early 21st century, the test industry has moved to the domestic, can be said to have completed the early stage of development in the Republic of Korea and Chinese Taiwan, labor-intensive IC test industry first transfer. The technology and capital intensive IC manufacturing industry is next, after the transfer will be 1-2 generations of technology. Knowledge-intensive IC design is generally difficult to transfer, the technology gap is significant, need to rely on independent development.

Project 4

Household Communication Control System Design

Task 1 Output a Simple Text Message from the RS232 Port

Objectives

After completing this task, you should be able to
- explain the pros and cons of parallel I/O and serial I/O;
- describe the operation of the UART modules;
- configure the UART modules to perform data transmission and reception.

Task Requirement

This task shows how we can interface our microcontroller to an external RS232 compatible device (e.g. an RS232 visual display unit, or COM1 or COM2 port of a PC) and send a text message to this device. The text message 'THIS IS AN RS232 TEST MESSAGE' is sent out continuously from the microcontroller. The frame format used in this project is 2,400 baud, 8 data bits, no parity, and 1 stop bit.

1.1 Serial Communication Basis and Serial Interface

Scan the QR code to view the teaching note of Serial Communication Basis and Serial Interface

Scan the QR code to watch the teaching video of Serial Communication Basis and Serial Interface

Serial Data Input/Output

Computers must be able to communicate with other computers in modern multiprocessor distributed systems. One cost-effective way to communicate is to send and receive data bits serially.

The 8051 has a serial data communication circuit that uses register SBUF to hold data. Register SCON controls data communication, register PCON controls data rates, and pins RXD (P3.0) and TXD (P3.1) connect to the serial data network.

SBUF has physically two registers. One is write only and is used to hold data to be transmitted out of the 8051 via TXD. The other one is read only and holds received data from external sources via RXD. Both mutually exclusive registers use address 99h.

There are four programmable modes for serial data communication that are chosen by setting the SMX bits in SCON. Baud rates are determined by the mode chosen. Table 4.1.1 and Table 4.1.2 show the bit assignments for SCON and PCON.

Table 4.1.1 The serial port control (SCON) special function register

Bit name	Bit position	Description
SM0	7	Serial port mode bit 0. Set/cleared by program to select mode
SM1	6	Serial port mode bit 1. Set/cleared by program to select mode <table><tr><th>SM0</th><th>SM1</th><th>Mode</th><th>Description</th></tr><tr><td>0</td><td>0</td><td>0</td><td>8-bit shift register; baud = $f/12$</td></tr><tr><td>0</td><td>1</td><td>1</td><td>10-bit UART; baud = variable</td></tr><tr><td>1</td><td>0</td><td>2</td><td>11-bit UART; baud = $f/32$ or $f/64$</td></tr><tr><td>1</td><td>1</td><td>3</td><td>11-bit UART; baud = variable</td></tr></table>
SM2	5	Multiprocessor communications bit. Set/cleared by program to enable multiprocessor communications in modes 2 and 3. When set to 1 an interrupt is generated if bit 9 of the received data is a 1; no interrupt is generated if bit 9 is a 0. If set to 1 for mode 1, no interrupt will be generated unless a valid stop bit is received. Clear to 0 if mode 0 is in use
REN	4	Receive enable bit. Set to 1 to enable reception; cleared to 0 to disable reception
TB8	3	Transmitted bit 8. Set/cleared by program in modes 2 and 3
RB8	2	Received bit 8. Bit 8 of received data in modes 2 and 3; stop bit in mode 1. Not used in mode 0
TI	1	Transmit interrupt flag. Set to one at the end of bit 7 time in mode 0, and at the beginning of the stop bit for other modes. Must be cleared by the program
RI	0	Receive interrupt flag. Set to one at the end of bit 7 time in mode 0, and halfway through the stop bit for other modes. Must be cleared by the program

Table 4.1.2 The power mode control (PCON) special function register

Bit name	Bit position	Description
SMOD	7	Serial baud rate modify bit. Set to 1 by program to double baud rate using timer 1 for modes 1, 2, and 3. Cleared to 0 by program to use timer 1 baud rate
—	4–6	Not implemented
GF1	3	General purpose user flag bit 1. Set/cleared by program
GF0	2	General purpose user flag bit 0. Set/cleared by program
PD	1	Power down bit. Set to 1 by program to enter power down configuration for CHMOS processors
IDL	0	Idle mode bit. Set to 1 by program to enter idle mode configuration for CHMOS processors. PCON is not bit addressable

Serial Data Interrupts

Serial data communication is a relatively slow process, occupying many milliseconds per data byte to accomplish. In order not to tie up valuable processor time, serial data flags are included in SCON to aid in efficient data transmission and reception notice that data transmission is under the complete control of the program, but reception of data is unpredictable and at random times that are beyond the control of the program.

The serial data flags in SCON, TI and RI, are set whenever a data byte is transmitted (TI) or received (RI). These fags are ORed together to produce an interrupt to the program. The program must read these flags to determine which causes the interrupt and then clear the fag. This is unlike the timer flags that are cleared automatically; it is the responsibility of the programmer to write routines that handle the serial data flags.

Data Transmission

Transmission of serial data bits begins anytime data is written to SBUf. TI is set to 1 when the data has been transmitted and signifies that SBUF is empty (for transmission purposes) and that another data byte can be sent. If the program fails to wait for the TI flag and overwrites SBUF while a previous data byte is in the process of being transmitted, the results will be unpredictable (a polite term for "garbage out").

Data Reception

Reception of serial data will begin if the receive enable bit (REN) in SCON is set to 1 for all modes. In addition, for mode 0 only, RI must be cleared to 0 too. Receiver interrupt flag RI is set after data has been received in all modes. Setting REN is the only direct program control that limits the reception of unexpected data; the requirement that RI is also set to 0 for mode 0 prevents the reception of new data until the program has dealt with the old data and reset RI.

Reception can begin in modes 1, 2, and 3 if RI is set when the serial stream of bits begins. RI must have been reset by the program before the last bit is received or the incoming data will be lost. Incoming data is not transferred to SBUF until the last data bit has been received so that the previous transmission can be read from SBUF while new data is being received.

Serial Data Transmission Modes

The 8051 designers have included four modes of serial data transmission that enable data communication to be done in a variety of ways and a multitude of baud rates modes are selected by the programmer by setting the mode bits SM0 and SM1 in SCON. Baud rates are fixed for mode 0 and variable, using timer 1 and the serial baud rate modify bit (SMOD) in PCON, for modes 1, 2, and 3.

Serial Data Mode 0 - Shift Register Mode

Setting bits SM0 and SM1 in SCON to 00b configures SBUF to receive or transmit eight data bits using pin RXD for both functions. Pin TXD is connected to the internal shift frequency pulse source to supply shift pulses to external circuits. The shift frequency, or baud rate, is fixed at 1/12 of the oscillator frequency, the same rate used by the timers when in the timer configuration. The TXD shift clock is a square wave that is low for machine cycle states S3-S4-S5 and high for S6-SI-S2. Figure 4.1.1 shows the timing for mode 0 shift register data transmission.

When transmitting, data is shifted out of RXD: The data changes on the falling edge of S6P2, or one clock pulses after the rising edge of the output TXD shift clock. The system designer must design the external circuitry that receives this transmitted data to receive the data reliably based on this timing.

Received data comes in on pin RXD and should be synchronized with the shift clock produced at TXD. Data is sampled on the falling edge of S5P2 and shifted in to SBUF on the rising edge of the shift clock.

Mode 0 is intended not for data communication between computers, but as a high speed serial data-collection method using discrete logic to achieve high data rates. The baud rate used in mode 0 will be much higher than standard rate for any reasonable oscillator frequency; for a 6 MHz crystal, the shift rate will be 500 KHz.

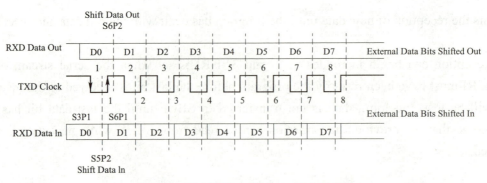

Figure 4.1.1   Shift register mode 0 timing

Serial Data Mode 1 — Standard UART

When SM0 and SM1 are set to 01b, SBUF becomes a 10-bit full-duplex receiver/transmitter that may receive and transmit data at the same time Pin RXD receives all data and pin TXD transmits all data. Figure 4.1.2 shows the standard UART data word.

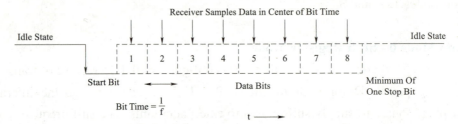

Figure 4.1.2   Standard UART data word

Transmitted data is sent as a start bit, eight data bits (Least Significant Bit, LSB first), and a stop bit. Interrupt flag TI is set once all ten bits have been sent. Each bit interval is the inverse of the baud rate frequency, and each bit is maintained high or low over that interval.

Received data is obtained in the same order; reception is triggered by the falling edge of the start bit and continues if the stop bit is true (0 level) halfway through the start bit interval. This is an anti-noise measure; if the reception circuit is triggered by noise on the transmission line, the check for a low after half a bit interval should limit false data reception.

Data bits are shifted into the receiver at the programmed baud rate, and the data word will be loaded to SBUF if the following conditions are true: RI must be 0, and mode bit SM2 is 0 or the stop bit is 1 (the normal state of stop bits). RI set to 0 implies that the program has read the previous data byte and is ready to receive the next; a normal stop bit will then complete the transfer of data to SBUF regardless of the state of SM2. SM2 set to 0 enables the reception of a byte with any stop bit state, a condition which is of limited use in this mode but very useful in modes 2 and 3. SM2 set to 1 forces reception of only "good" stop bits, an anti-noise safeguard.

Of the original ten bits, the start bit is discarded, the eight data bits go to SBUF, and the stop bit is saved in bit RB8 of SCON. RI is set to 1, indicating a new data byte has been received.

If RI is found to be set at the end of the reception, indicating that the previously received data byte has not been read by the program, or if the other conditions listed are not true, the new data will not be loaded and will be lost.

Mode 1 Baud Rates

Timer 1 is used to generate the baud rate for mode 1 by using the overflow flag of the timer to determine the baud frequency. Typically, timer 1 is used in timer mode 2 as an auto-load 8-bit timer that generates the baud frequency:

$$f_{baud} = \frac{2^{SMOD}}{32d} \times \frac{f_{osc}}{12d \times [256d - (TH1)]}$$

SMOD is the control bit in PCON and can be 0 or 1, which raises the two in the equation to a value of 1 or 2.

If timer 1 is not run in timer mode 2, then the baud rate is

$$f_{baud} = \frac{2^{SMOD}}{32d} \times (T_1 \text{ overflow frequency})$$

and timer 1 can be run using the internal clock or as a counter that receives clock pulses from any external source via pin T1.

The oscillator frequency is chosen to help generate both standard and nonstandard baud rates. If standard baud rates are desired, then an 11.059,2 MHz crystal could be selected.

To get a standard rate of 9,600 Hz then, the setting of TH1 may be found as follows:

$$TH1 = 256d - \left(\frac{2^0}{32d} \times \frac{11.059,2 \times 10^6}{12 \times 9,600d} \right) = 253.000,0d = 0\text{FDh}$$

if SMOD is cleared to 0.

Serial Data Mode 2 - Multiprocessor Mode

Mode 2 is similar to mode 1 except 11 bits are transmitted: a start bit, nine data bits and a stop bit, as is shown in Figure 4.1.3. The ninth data bit is gotten from bit TB8 in SCON when transmitted and stored in bit RB8 of SCON when data is received. Both the start and stop bits are discarded.

The baud rate is programmed as follows:

$$f_{baud} = \frac{2^{SMOD}}{64d} \times f_{osc}$$

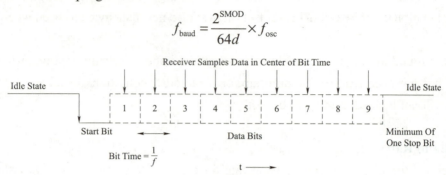

Figure 4.1.3 Multiprocessor data word

Here, as in the case for mode 0, the baud rate is much higher than standard communication rates. This high data rate is needed in many multi-processor applications. Data can be collected quickly from an extensive network of communicating microcontrollers if high baud rates are employed.

The conditions for setting RI for mode 2 are similar to mode 1: RI must be 0 before the last bit is received, and SM2 must be 0 or the ninth data bit must be 1. Setting RI based upon the state of SM2 in the receiving 8051 and the state of bit 9 in the transmitted message makes multiprocessing possible by enabling some receivers to be interrupted by certain messages, while other receivers ignore those messages. Only those 8051's that have SM2 set too will be interrupted by received data which has the ninth data bit set to 0; those with SM2 set to 1 will not be interrupted by messages with data bit 9 at 0. All receivers will be interrupted by data words that have the ninth data bit set to 1; the state of SM2 will not block reception of such messages.

This scheme allows the transmitting computer to "talk" to selected receiving computers without interrupting other receiving computers. Receiving computers can be commanded by the "talker" to "listen" or "deafen" by transmitting coded byte(s) with the ninth data bit set to 1. The 1 in data bit 9 interrupts all receivers, instructing those that are programmed to respond to the coded byte(s) to program the state of SM2 in their respective SCON registers. Selected listeners then respond to the bit 9 set to 0 messages, while other receivers ignore these messages. The talker can change the mix of listeners by transmitting bit 9 set to 1 messages that instruct new listeners to set SM2 to 0, while others are instructed to set SM2 to 1.

Serial Data Mode 3

Mode 3 is identical to mode 2 except that the baud rate is determined exactly as in mode 1, using timer 1 to generate communication frequencies.

1.2 RS232 Serial Communication

Scan the QR code to view the teaching note of RS232 Serial Communication

Scan the QR code to watch the teaching video of RS232 Serial Communication

RS232 is a serial communications standard which enables data to be transferred in serial form between two devices. Data is transmitted and received in serial "bit stream" from one point to another. Standard RS232 is suitable for data transfer to about 50m, although special low-loss cables can be used for extended distance operation. Four parameters specify an RS232 link between two devices. These are baud rate, data width, parity, and the stop bits, and are described below:

- Baud rate: The baud rate (bits per second) determines how much information is transferred over a given time interval. A baud rate can usually be selected between 110 and 76,800 baud, e.g., a baud rate of 9,600 corresponds to 9,600 bits per second.
- Data width: The data width can be either 8 bits or 7 bits depending upon the nature of the data being transferred.
- Parity: The parity bit is used to check the correctness of the transmitted or received data. Parity can either be even, odd, or no parity bit can be specified at all.
- Stop bit: The stop bit is used as the terminator bit and it is possible to specify either one or two stop bits.

Serial data is transmitted and received in frames where a frame consists of the following:

- 1 start bit;
- 7 or 8 data bits;
- optional parity bit;
- 1 stop bit.

In many applications 10 bits are used to specify an RS232 frame, consisting of 1 start bit, 8 data bits, no parity bit, and 1 stop bit. For example, character 'A' has the ASCII bit pattern

'01000001' and is transmitted as is shown in Figure 4.1.4 with 1 start bit, 8 data bits, no parity, and 1 stop bit. The data is transmitted least significant bit first.

When 10 bits are used to specify the frame length, the time taken to transmit or receive each bit can be found from the baud rate used. Table 4.1.3 gives the time taken for each bit to be transmitted or received for most commonly used baud rates.

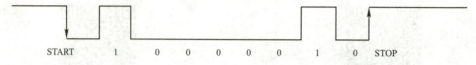

Figure 4.1.4 Transmitting character 'A' (bit pattern 01000001)

Table 4.1.3 Bit times for most commonly baud rates

Baud rate	Bit time
300	3.33ms
600	1.66ms
1,200	833μs
2,400	416μs
4,800	208μs
9,600	104μs
19,200	52μs

RS232 Connectors

As is shown in Figure 4.1.5, two types of connectors are used for RS232 communications. These are the 25-way D-type connectors (known as DB25) and the 9-pin D-type connectors (also known as DB9). Table 4.1.4 lists the most commonly used signal names for both DB9 and DB25 type connectors. The used signals are:

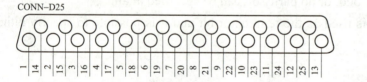

Figure 4.1.5 RS232 connectors

Table 4.1.4 Commonly used RS232 signals

Description	Signal	9 pin	25 pin
Carrier detect	CD	1	8
Receive data	RD	2	3
Transmit data	TD	3	2
Data terminal ready	DTR	4	20
Signal ground	SG	5	7

continued

Description	Signal	9 pin	25 pin
Data set ready	DSR	6	6
Request to send	RTS	7	4
Clear to send	CTS	8	5
Ring indicator	RI	9	22

SG: signal ground. This pin is used in all RS232 cables.

RD: received data. Data is received at this pin. This pin is used in all two-way communications.

TD: transmit data. Data is sent out from this pin. This pin is used in all two-way communications.

RTS: request to send. This signal is asserted when the device requests data to be sent.

CTS: clear to send. This signal is asserted when the device is ready to accept data.

DTR: data terminal ready. This signal is asserted to indicate that the device is ready.

DSR: data set ready. This signal indicates, by the device at the other end, that it is ready.

CD: carrier detect. This signal indicates that a carrier signal has been detected by a modem connected to the line.

RI: ring indicator. When the local modem receives a ringing call signal from the switchboard, the signal is valid and notifies the terminal that it has been called.

In some RS232 applications it is sufficient to use only the pins SG, RD, and TD. Also, in some applications (e.g. when two similar devices are connected together) it is necessary to twist pins RD and TD so that the transmit pin of one device is connected to the receive pin of the other device and vice versa.

RS232 Signal Levels

RS232 is bi-polar and a voltage of +3 to +12 V indicates an ON state (or SPACE), while a voltage of −3 to +12 V indicates an OFF state (or MARK). In practice, the ON and OFF states can be achieved with lower voltages.

Standard TTL logic devices, including the 89C51 microcontroller, operate with TTL logic levels between the voltages of 0 and +5 V. Voltage level converter ICs are used to convert between the TTL and RS232 voltage levels. One such popular IC is the MAX232, manufactured by MAXIM, and operators with +5 V supply. The MAX232 is a 16-pin DIL chip incorporating two receivers and two transmitters (see Figure 4.1.6) and the device requires four external capacitors for proper operation.

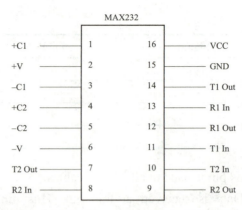

Figure 4.1.6 Pin configuration of MAX232

The 89C51 microcontroller can output TTL level RS232 signals from its TXD (or pin P3.1) pin and it can receive TTL level RS232 signals from its RXD (or pin P3.0) pin. The microcontroller can be connected to external RS232 compatible equipments via a MAX232 type voltage converter IC.

Controlling the RS232 Port

Before the serial port can be used, it is necessary to set various registers:

SCON: This is the serial port control register. It should be set to hexadecimal 0x50 for 8-bit data mode.

TMOD: This register controls the timers for baud rate generation and it should be set to hexadecimal 0x20 to enable timer 1 to operate in 8-bit auto-reload mode.

TH1: This register should be loaded with a constant so that the required baud rate can be generated. Table 4.1.5 shows the values to be loaded into TH1 and the corresponding baud rates for two different clock rates.

TR1: This register starts/stops the timer and it should be set to 1 to start timer 1.

TI: This register should be set to 1 to indicate ready to transmit.

Table 4.1.5 TH1 values for different baud rates

Baud rate	Clock	SMOD	TH1 value	Error
9,600	12.000 MHz	1	0xF9	7%
4,800	12.000 MHz	0	0xF9	7%
2,400	12.000 MHz	0	0xF3	0.16%
1,200	12.000 MHz	0	0xE6	0.16%
9,600	11.059 MHz	0	0xFD	0
4,800	11.059 MHz	0	0xFA	0
2,400	11.059 MHz	0	0xF4	0
1,200	11.059 MHz	0	0xE8	0

Note that register SMOD should be set to 1 when we require 9,600 baud at 12 MHz clock rate. SMOD is set to 0 at reset time.

For example, the following function shows how we can initialize the serial port for 2,400 baud operation:

```
void serial_init()
{
    SCON=0x50;
    TMOD=0x20;
    TH1=0xF3;
    TR1=1;
    TI=1;
}
```

Circuit Diagram

The block diagram is shown in Figure 4.1.7. The TXD pin of the microcontroller is connected to the MAX232 Maxim voltage converter IC and the output of this IC can be connected to the input of a COM1 (or COM2) port of a PC, or to the input of an RS232 visual display unit. Similarly, the output of the external RS232 device is connected to the RXD input of the micro controller via the MAX232 IC. A terminal emulation software can be activated on the PC to receive and display any data arriving at its serial port.

The complete circuit diagram of this project is shown in Figure 4.1.8. Pin P3.1 of the microcontroller (TXD) is connected to pin 10 of the MAX232 converter IC. Pin 7 of this IC is connected to the external RS232 compatible serial device which is to receive and display our text message. Similarly, the output of the RS232 device is connected to pin 8 input of the MAX232 IC and pin 9 output of this IC is connected to pin 2 (RXD) serial input of the microcontroller. Correct operation of MAX232 requires four external capacitors to be connected, as is shown in Figure 4.1.8.

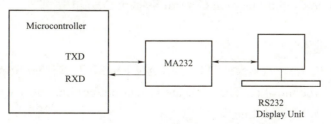

Figure 4.1.7   Block diagram

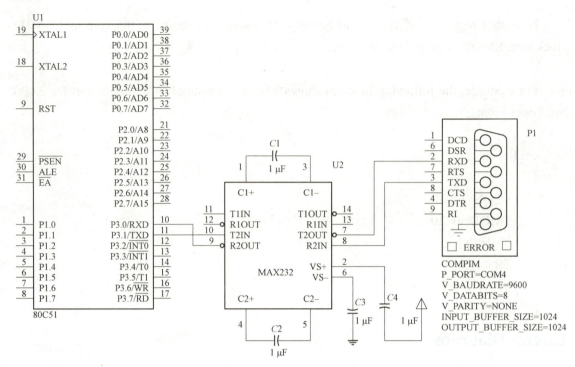

Figure 4.1.8   Circuit diagram

Program Description

The program initializes the RS232 port of the microcontroller and then sends a test message to the port.

The program listing is given as following. Notice that the standard input/output library 'stdio.h' is included at the top of the program. The main program calls function serial_init which initializes the RS232 port to 2,400 baud and enables transmissions. Standard C function printf is used to send the serial data to the RS232 port. A carriage return and line feed pair ('\n') are sent after each output.

```
/*********************************************************************
PROJECT: Household Communication Control System Design
FILE: PROJ4_1(1).C
PROCESSOR: AT89C51
This project sends the text message: 'THIS IS AN RS232 TEST MESSAGE' to the RS232 serial port of the microcontroller. The message is sent out continuously with a carriage return and line feed at the end of each line.
/*********************************************************************
```

```c
#include <stdio.h>
#include <AT89C51.h>
/* Function to initialize the RS232 serial port */
void serial_init()
{
    SCON=0x50;          /* setup for 8-bit data */
    TMOD=0x20;          /* setup timer 1 for auto-reload */
    TH1=0xF3;           /* setup for 2400 baud */
    TR1=1;              /* turn on timer 1 */
    TI=1;               /* indicate ready to transmit */
}

/* Start of main program */
main()
{
    serial_init();      /*initialize serial port*/
    for(;;)             /*start of loop*/
    {
        printf('THIS IS AN RS232 TEST MESSAGE\n');
    }
}
```

It is important to notice that this simple program occupies about 1,094 Bytes in the memory of the microcontroller. This is because the printf function is a complex library function and is implemented in a large number of instructions. A simple function can be developed to emulate some of the functionalities of printf so that the output operations can be performed with less memory as described below.

A Simple Serial Output Function

It was shown in the previous example that using the built-in printf function causes a large part of this memory to be used, leaving little space for other operations. The following code shows a program listing that performs serial output functions without using the printf function and the complete program occupies about 400 bytes of memory. In this program, the serial transmit register of the microcontroller (SBUF) is used to send out data directly. Function send_serial transmits a null-terminated string to the RS232 port of the microcontroller. The program waits until the transmit register is empty (TI = 1) before sending out the next character. In this example, the string 'ANOTHER TEST' is output continuously. Notice that calling this function with variable crlf causes a carriage return and line feed to be output at the end of the test message.

```
/****************************************************************
PROJECT: Independent Button Access Control System Design
FILE: PROJ4_1(2).C
PROCESSOR: AT89C51
```
This project sends the text message: 'ANOTHER TEST' to the RS232 serial port of the microcontroller. The message is sent out continuously with a carriage return and line feed at the end of each line.
This program does not use the built-in function printf. The program occupies about 400 bytes of memory.
```
*****************************************************************/
#include <stdio.h>
#include <AT89C51.h>
/* Function to initialize the RS232 serial port */
void serial_init()
{
    SCON=0x50;          /* setup for 8-bit data */
    TMOD=0x20;          /* setup timer 1 for auto-reload */
    TH1=0xF3;           /* setup for 2400 baud */
    TR1=1;              /* turn on timer 1 */
    TI=1;               /* indicate ready to transmit */
}

/* This function displays a null-terminated string on the RS232 port */
void send_serial(unsigned char *s)
{
    while(*s != 0x0)
    {
        SBUF=*s;        /*send out the character*/
        while(! TI)     /*wait until sent*/
        {
        }
        TI=0;
        s++;            /*get the next character*/
    }
}
```

```c
/* Start of main program */
main()
{
    unsigned char crlf[ ]={0x0D,0x0A,0x0};   /*carriage return, line feed*/
    serial_init();                            /*initialize serial port*/
    for(;;)                                   /*Start of loop*/
    {
        send_serial(`ANOTHER TEST');
        send_serial(crlf);
    }
}
```

Quiz

1. When single chip microcomputer carries out multi-machine communication, the working mode of serial interface should be selected as (   ).
   A) mode 0
   B) mode 1
   C) mode 2
   D) mode 2 or mode 3

2. When sending serial data, after sending a frame of data, the TI flag will (   )?
   A) automatic reset
   B) hardware reset
   C) software reset
   D) both software and hardware are acceptable

3. The characteristics of serial communication are (   ).
   A) long transmission distance and relatively fast transmission speed
   B) short transmission distance and relatively fast transmission speed
   C) long transmission distance and relatively slow transmission speed
   D) short transmission distance and relatively slow transmission speed

4. The asynchronous serial communication mode with frame format of 1 start bit, 8 data bits and 1 stop bit is (   ).
   A) mode 0   B) mode 1   C) mode 2   D) mode 3

5. In serial communication, if data can be sent from station A to station B at the same time, and from station B to station A, this communication mode is (   ).
   A) simplex   B) half duplex   C) full duplex   D) multiplex

MCU Practical Technology

Scan the QR code to view the answer of quiz

Summary

The communication between computers or between computers and peripherals can be divided into parallel communication and serial communication.

MCU internal with a full duplex asynchronous communication interface, the serial port has four working modes, the special rate and format can be programmed. The frame format has 10 and 11 bits. The band rate of working mode 0 and working mode 2 are fixed, while the baud rate of working mode 1 and working mode 3 are variable, which is determined by the overflow rate of timer T1.

Between MCU and MCU and between MCU and PC can communicate, the control program design usually uses two methods: query method and interrupt method.

Task 2 Input/Output Example Using the RS232 Port

Objectives

After completing this task, you should be able to
- call UART functions to read data and transmit data;
- use UART module to interface with shift registers.

Task Requirement

This task is an example of using both the input and the output serial data routines. Microprocessor A is used to send data. Microprocessor B will display the received data in 7-segment digital tube.

Scan the QR code to view the teaching note of Serial Communication Operation Process

Scan the QR code to watch the teaching video of Serial Communication Operation Process

Circuit Diagram

The circuit diagram is shown in Figure 4.2.1.

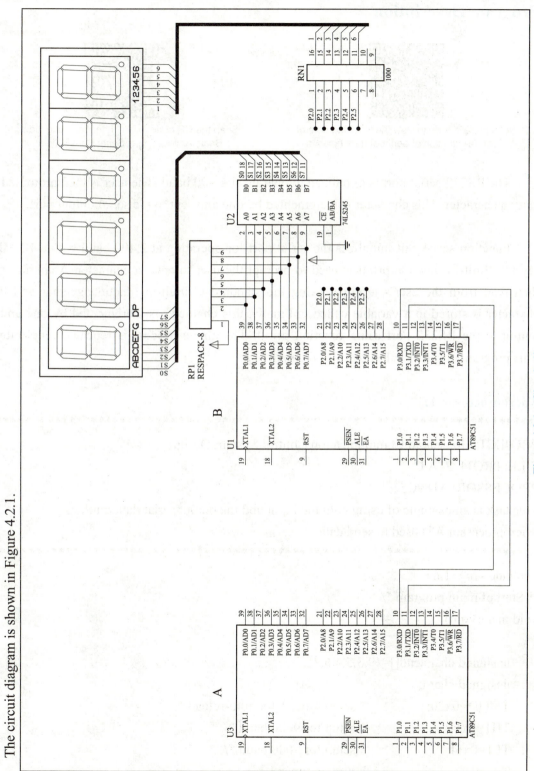

Figure 4.2.1 Circuit diagram

Program Description

Scan the QR code to view the teaching note of Basic Application of Serial Communication

Scan the QR code to watch the teaching video of Basic Application of Serial Communication

The RS232 serial port is initialized to operate at 2,400 baud. The user is then prompted to enter a character. This character is incremented by one and sent to the serial output port.

Function serial_init initializes the serial port for operation at 2,400 baud with a 12 MHz crystal. Built-in function printf is used to prompt the user to enter a character. A character is then read from the user's terminal using the standard C built-in function getchar and this character is stored in a variable called c. Finally, this character is incremented by one and is output to the RS232 port using function printf. The above process is repeated indefinitely. This program occupies 1,164 bytes of the memory.

Microprocessor A:

```
/************************************************************
PROJECT: Household Communication Control System Design
FILE: PROJ4_2_A.C
PROCESSOR: AT89C51
This task is an example of using both the input and the output serial data routines.
Microprocessor A is used to send data.
*************************************************************/
#include <reg51.h>
/* Start of main program */
void main(void)
{
    unsigned char send[]={9,3,5,4,6,7};
    unsigned char i;
    TMOD=0x20;            /* setup timer 1 for auto-reload */
    TH1=0xf4;             /* setup for 9600 baud */
    TL1=0xf4;             /* setup for 9600 baud */
    TR1=1;                /* turn on timer 1 */
```

```
    TI=1;                    /* indicate ready to transmit */
    SCON=0x40;
    PCON=0x00;

    for(i=0;i<6;i++)         /*send the characters*/
    {
        SBUF=send[i];
        while(TI==0);
        TI=0;
    }
    while(1);
}
```

Microprocessor B:

```
/************************************************************************
PROJECT: Household Communication Control System Design
FILE: PROJ4_2_B.C
PROCESSOR: AT89C51
This task is an example of using both the input and the output serial data routines.
Microprocessor B will display the received data in 7-segment digital tube.
************************************************************************/
#include <reg51.h>
unsigned char code tab[]={0xc0,0xf9,0xa4,0xb0,0x99,0x92,0x82,0xf8,0x80,0x90};
unsigned char buffer[]={0x00,0x00,0x00,0x00,0x00,0x00};

void disp(void);

/* Start of main program */
void main(void)
{
    unsigned char i;
    TMOD=0x20;               /* setup timer 1 for auto-reload */
    TL1=0xf4;                /* setup for 9600 baud */
    TH1=0xf4;                /* setup for 9600 baud */
    TR1=1;                   /* turn on timer 1 */
    SCON=0x40;
    REN=1;
```

```
        ES=1;              /*Enable the serial interrupt*/
        EA=1;
        i=0;

        while(1) disp();
}

/*The digital tube display function*/
void disp(void)
{
    unsigned char w,i,j;
    w=0x01;
    for(i=0;i<6;i++)
    {
        P0=0x00;
        P0=~tab[buffer[i]];
        P2=~w;
        for(j=0;j<100;j++);
        w<<=1;
    }
}

/*The serial interrupt function*/
void serial () interrupt 4
{
    EA=0;
    RI=0;
    buffer[i]=SBUF;   /*receive the characters*/
    i++;
    if(i==6) i=0;
    EA=1;
}
```

Input/Output Without Using the Built-in Functions

The above program uses the standard C built-in functions printf and getchar. As a result the program is big. An example program above which does not use these built-in functions and thus occupies much less space in memory.

Project 4 Household Communication Control System Design

Function serial_init is the same as before but note that the serial port interrupts are enabled (EA=1 and ES=1). Function send_serial sends a null-terminated string to the serial output port. Similarly, function send_1_char sends a single character to the serial port. Serial data is read in via the serial port interrupt service routine (serial). Whenever a character is transmitted or received, the interrupt service routine is activated automatically. The interrupt number of the serial port is 4. Here, the receive interrupt register (RI) is checked and a character is assumed to be received from the serial port if RI is non-zero. The received character is copied from SBUF to a variable called received_character.

Quiz

1. When parallel I/O ports are extended using a serial port, the serial port should be set to work (　　).
   A) 0
   B) 1
   C) 2
   D) 3

2. The register that controls how the serial port works is (　　).
   A) TCON
   B) PCON
   C) SCON
   D) TMOD

3. The serial port transmits (　　) characters every time.
   A) 1 bit
   B) 1 string
   C) 1 frame
   D) 1 baud

4. There are several modes of operation of serial port. (　　)
   A) 1
   B) 2
   C) 3
   D) 4

5. The baud rate of serial port mode 1 is (　　).
   A) fixed, $f_{osc}/32$

B) fixed, $f_{osc}/16$
C) variable, through the timer/counter T1 overflow rate set
D) fixed, $f_{osc}/64$

Scan the QR code to view the answer of quiz

Summary

The key contents to be mastered in this project are as follows:

1) Basic knowledge of serial communication;
2) Serial port structure, working mode and baud rate setting;
3) Dual-machine communication between MCU;
4) Communication between MCU and PC;
5) Extension of parallel I/O port.

Extended Reading

Top 10 IC design companies in China

In recent years, China's chip design industry has maintained a trend of rapid growth driven by factors such as increasing self-sufficiency, policy support, upgrading specifications and innovative applications. The chip design industry's sales revenue rose from 132.5 billion yuan in 2015 to 294.77 billion yuan in 2019, as data showed. It is expected that the market size of China's chip industry will exceed 350 billion yuan in 2020, and the chip design industry has become one of the most dynamic fields in the domestic semiconductor industry.

The well-known research institutions IC Insights and issued the international strategy of science and technology development related data, published in 2019—China top 10 ranking of IC design enterprises, respectively in turn, Hisilicon, Unigroup, Omnivision, Bitmain , ZTE Microelectronics, Huada Integrated Circuit, Nari Smart Chip, ISSI, Zhaoyi Innovation and Datang Semiconductor.

1. HiSilicon

HiSilicon has grown into top 1 of China, and one of the world's top 10 IC design companies, which owns five series chips. They are respectively used for smartphone household Kirin Kunpeng series CPU chip, for data center server CPU chip, used in the scenario of artificial intelligence AI chipset rise series of SoC, for 4G and 5G connection chip (base station plough, terminal chip barone) and other special video monitoring, set-top box, smart TV, Internet and other chips. HiSilicon is the most anticipated company in the field of chip design. But because of the policy restrictions, high-end Kirin chips may no longer be able to produce, HiSilicon future development will be no small impact.

2. Unigroup

Unisplendour, the chip design company of Unigroup, is the second largest IC chip design company in China after HiSilicon, and has already entered the first echelon of 5G chips in the world. At present, The chips developed by Unisplendour in the 5G field have been attracting the attention of many manufacturers and netizens.

3. Omnivision

Founded in the United States in 1995, Omnivision is a well-known semiconductor company in the United States. It mainly designs and sells high-performance semiconductor image sensors, and is known as the world's leading three major image sensor suppliers together with SONY of Japan and Samsung of the Republic of Korea. In May 2015, it was directly acquired by a domestic consortium for 1.9 billion US dollars, and then acquired by Weil Shares for 15.3 billion yuan in May 2018.

4. Bitmain

Bitmain belongs to the brand of Beijing Bitmain Technology Limited Company, founded in 2013, it is mainly engaged in the design and development of high-speed and low-power customized chips, which are mainly applied in finance, high-performance computing HPC, machine learning algorithm, AI artificial intelligence and other fields. Now its sales and services cover more than 100 countries around the world.

5. ZTE Microelectronics

Founded in 2003, ZTE Microelectronics is a chip manufacturer under ZTE corporation. At present, ZTE's complex SoC chip design capability has reached the international leading level, with the ability to customize the whole process from front-end design, back-end design to package and test, which can provide the overall chip solutions. From the perspective of technological process, in 2018, the shipments of advanced technology chips with 28nm or below accounted for 84%, and the technological level of products under development has reached 7nm, and the process of 5nm has been introduced simultaneously, which is also the world's leading level. However, ZTE in terminal chips, server chips and other weaknesses, there is still room for further improvement.

6. Huada Integrated Circuit

China Huada Integrated Circuit Design Limited Company was founded in October 2003, which is a state-owned large-scale integrated circuit design company, on the basis of seven of the country's investment in "909" design company consolidated to form a group of integrated circuit design company.

7. Nari Smart Chip

Nari Smart Chip is directly under the State Grid Corporation Nanrui Group, an IC leading enterprise in the power industry.

8. ISSI

ISSI, a former Nasdaq-listed company, was taken private in late 2015 for $780 million by Beijing Si Cheng. Main types of high-performance DRAM, SRAM, FLASH memory chips and ANALOG chip research and development and sales. In 2019, Beijing Junzheng acquired Beijing Si Cheng for 7.2 billion yuan, bringing ISSI under its control.

9. Zhaoyi Innovation

Zhaoyi Innovation is a leading provider of memory technology and IC solutions. Mega Innovation currently has three main businesses: memory chips, MCU and sensors.The memory chip business is the core business of the company. NOR Flash is the core product of the company. NOR Flash occupies the first place in the domestic market and is also one of the top three NOR Flash suppliers in the world.

10. Datang Semiconductor

Datang Semiconductor Design Limited Company is a wholly-owned subsidiary of Datang Telecom Technology Limited Company. Specific products include mobile terminal chips, smart card security chips, new energy vehicles and hybrid electric vehicle power management driver chips.

Glossary

ADC Analogue-to-digital converter. A device that converts analogue signals to a digital form for use by a computer.

Algorithm A fixed step-by-step procedure for finding a solution to a problem.

ANSI American National Standards Institute.

Architecture The arrangement of functional blocks in a computer system.

ASCII American Standard Code for Information Interchange. A widely used code in which alphanumeric characters and certain other special characters are represented by unique 7-bit binary numbers. For example, the ASCII code of the letter 'A' is 65.

Assembler A software that translates symbolically represented instructions into their binary equivalents.

Assembly language A source language that is made up of the symbolic machine language statements. Assembly language is very efficient since there is a one-to-one correspondence with the instruction formats and data formats of the computer.

BASIC Beginners All-purpose Symbolic Instruction Code. A high-level programming language commonly used in personal computers. BASIC is usually an interpreted language.

Baud The unit of data transmission speed. Baud is often equated to the number of serial bits transferred per second.

Baud rate Measurement of data flow in a serial communication system. Baud rate is typically equal to bits per second. Some typical baud rates are 9,600, 4,800, 2,400 and so on.

BCD Binary Coded Decimal. A code in which each decimal digit is binary coded into 4-bit words. By representing binary numbers in this form, it is readily possible to display and print numbers.

Bi-directional port An interface port that can be used to transfer data in either direction.

Binary The representation of numbers in a base two system.

Bit A single binary digit.

Byte A group of 8 binary digits.

Chip A small rectangle of silicon on which an integrated circuit is fabricated.

Clock A circuit generating regular timing signals for a digital logic system. In microcomputer systems clocks are usually generated by using crystal devices. A typical clock frequency is 12 MHz.

CMOS Complementary Metal Oxide Semiconductor. A family of integrated circuits that offers extremely high packing density and low power.

Compiler A program designed to translate high-level languages into machine code.

Counter A register or a memory location used to record numbers of events as they occur.

CRT Cathode Ray Tube. A display screen.

Cycle time The time required to access a memory location or to carry out an operation in a computer system.

DAC Digital-to-analogue converter. A device that converts digital signals into analogue form.

Decimal system Base 10 numbering system.

Development system Equipment used to develop microprocessor and microcomputer-based software and hardware projects.

Dot matrix Method of printing or displaying characters in which each character is formed by a rectangular array of dots to give the required shape.

EAROM Electrically Alterable Read Only Memory. In this type of memory part or all of the data can be erased and rewritten by applying electrical signals.

Edge triggered Circuit action initiated by the change of a signal. An edge could be the change of a signal from 0 to 1 or from 1 to 0.

Emulator Software or hardware system that duplicates the actions of a microprocessor or a microcomputer system.

EPROM Erasable Programmable Read Only Memory. This type of memory can be erased by exposure to ultraviolet light and then reprogrammed using a programmer.

Execute To perform a specified operational sequence in a program.

File Logical collection of data.

Flow chart Graphical representation of the operation of a program.

Gate A logic circuit having one or more inputs and a single output. For example, NAND gate.

Half duplex A two-way communication system that permits communication in one direction at a time.

Hardware The physical parts or electronic circuitry of a computer system.

Hexadecimal Base 16 numbering system. In hexadecimal notation, numbers are represented by the digits 0±9 and the characters A±F. For example, decimal number 165 is represented as A5.

High-level language Programming language in which each instruction or statement corresponds to several machine code instructions. Some high-level languages are BASIC,

FORTRAN, C, PASCAL and so on.

Input device An external device connected to the input port of a computer. For example, a keyboard is an input device.

Input port Part of a computer that allows external signals to be passed into it. Microcomputer input ports are usually 8 bits wide.

I/O Short for Input Output.

Input/Output The hardware within the computer that connects the computer to external peripherals and devices.

Instruction cycle The process of fetching an instruction from memory and executing it.

Instruction set The complete set of instructions of a microprocessor or a microcomputer.

Interface To interconnect a computer to external devices and circuits.

Interrupt An external or internal event that suspends the normal program flow within a computer and causes entry into a special interrupt program (also called the interrupt service routine). For example, an external interrupt could be generated when a button is pressed. An internal interrupt could be generated when a timer reaches a certain value.

Interrupt vector Reserved memory locations where a program jumps when an interrupt is detected.

ISR Interrupt Service Routine. A program that is entered when an external or an internal interrupt occurs. Interrupt service routines are usually high priority routines.

K Multiplier for 1,024. For example, 1 kbyte is 1,024 bytes.

Language A prescribed set of characters and symbols which is used to convey a program to a computer.

LCD Liquid Crystal Display. A low-powered display that operates on the principle of reflecting incident light. An LCD does not itself emit light. There are many varieties of LCDs. For example, numeric, alphanumeric, or graphical.

LED Light Emitting Diode. A semiconductor device that emits a light when a current is passed in the forward direction. There are many colours of LEDs. For example, red, yellow, green, and white.

Level triggered Circuit action initiated by the presence of a signal.

Logic levels Voltage levels representing the two logical states (0 and 1) of a digital signal. Logic HIGH is also called state 1 and logic LOW is called state 0.

LSD Least Significant Digit. The right-most digit. For example, the LSD of number 123 is 3.

Machine code The lowest level in which programs are written. Machine code is usually written in hexadecimal.

Microcomputer General-purpose computer using a microprocessor as the CPU. A microcomputer consists of a microprocessor, a memory, and input/output ports.

Microprocessor A single large-scale integrated circuit which performs the functions of a CPU.

Mnemonic A programming shorthand using letters, numbers, and symbols adopted by each manufacturer to represent the instruction set of a micro processor.

MSD Most Significant Digit. The left-most digit of a number. For example, the MSD of number 123 is 1.

Nibble A group of 4 binary bits.

NMOS Negative channel Metal Oxide Semiconductor. A device based on an n-channel field-effect transistor cell.

Non-volatile memory A semiconductor memory type that holds data even if power has been disconnected.

Octal Representation of numbers in base 8.

Op-code Operation Code. That is part of an instruction which species the function to be performed.

Output device An external device connected to the output port of a computer. For example, a printer is an output device.

Output port Part of a computer that allows electrical signals to pass outside it. Microcomputer output ports are usually 8 bits wide.

Parity A binary digit added to the end of an array of bits to make the sum of all ones either odd or even. Parity is a method of checking the accuracy of transmitted or received binary data.

PDL Program Description Language. Representation of the control and data flow in a program using simple English-like sentences.

PEROM Flash Programmable and Erasable Read Only Memory. This type of memory can be erased and reprogrammed using electrical signals only, i.e., there is no need to use an ultraviolet light source to erase the memory.

Port An electrical logic circuit that is a signal input or output access point of a computer.

Programmed I/O The control of data flow in and out of a computer under software control.

PROM Programmable Read Only Memory. A type of semiconductor memory which can be programmed by the user using a special piece of equipment called a PROM programmer (or PROM blower).

Pull-up resistor A resistor connected to the output of an open collector (or open drain) transistor of a gate in order to load the output.

RAM Random Access Memory, also called read/write memory. Data in RAM is said to be volatile and it is present only as long as the chips have power supplied to them. When the

Glossary

power is cut off, all information disappears.

Register   A storage element in a computer. A register is usually 8 bits wide in most microprocessors and microcomputers.

ROM   Read Only Memory. A type of semiconductor memory that is read only.

RS232   An internationally recognized specification for serial data transfer between two devices.

Serial   Information transfer on a single wire where each bit is transferred sequentially with a time delay in between.

Software   Program.

Start bit   The first bit sent in a serial communication. There is only one start bit in a frame of serial communication.

Stop bit   The last bit sent in a serial communication. There can be one or two stop bits per frame of a serial communication.

Syntax   The rules governing the structure of a programming language.

Transducer   A device that converts a measurable quantity into an electronic signal. For example, a temperature transducer gives out an electrical signal which may be proportional to the temperature.

TTL   Transistor Transistor Logic. A kind of bipolar digital circuit.

UART   Universal Asynchronous Receiver Transmitter. This is a semiconductor chip that converts parallel data into serial form and serial data into parallel form. A UART is used in RS232 type serial communication.

USART   Synchronous version of UART.

UV   Ultraviolet light. Used to erase EPROM memories.

VDU   Visual Display Unit.

Word   A group of 16 binary digit.

References

[1] Intel. MCS51 Microcontroller Family User's Manual, 1994.

[2] Dogan Ibrahim. Microcontroller Projects in C for the 8051[M]. Oxford: Newnes, 2000.

[3] Kenneth J. Ayala. The 8051 Microcontroller [M]. St. Paul: West Publishing Company, 1991.

[4] Brian W. Kernighan. The C Programming Language [M]. New Jersey: Pearson Education, 2004.

[5] Stephen G. Kochan. Programming in C—A Complete Introduction to the C Programming Language [M]. New Jersey: Pearson Education, 2004.

[6] Zhang Wenqian. The C8051 Microcontroller—A Bilingual Approach [M]. Wuhan: Huazhong University of Science & Technology Press, 2019.